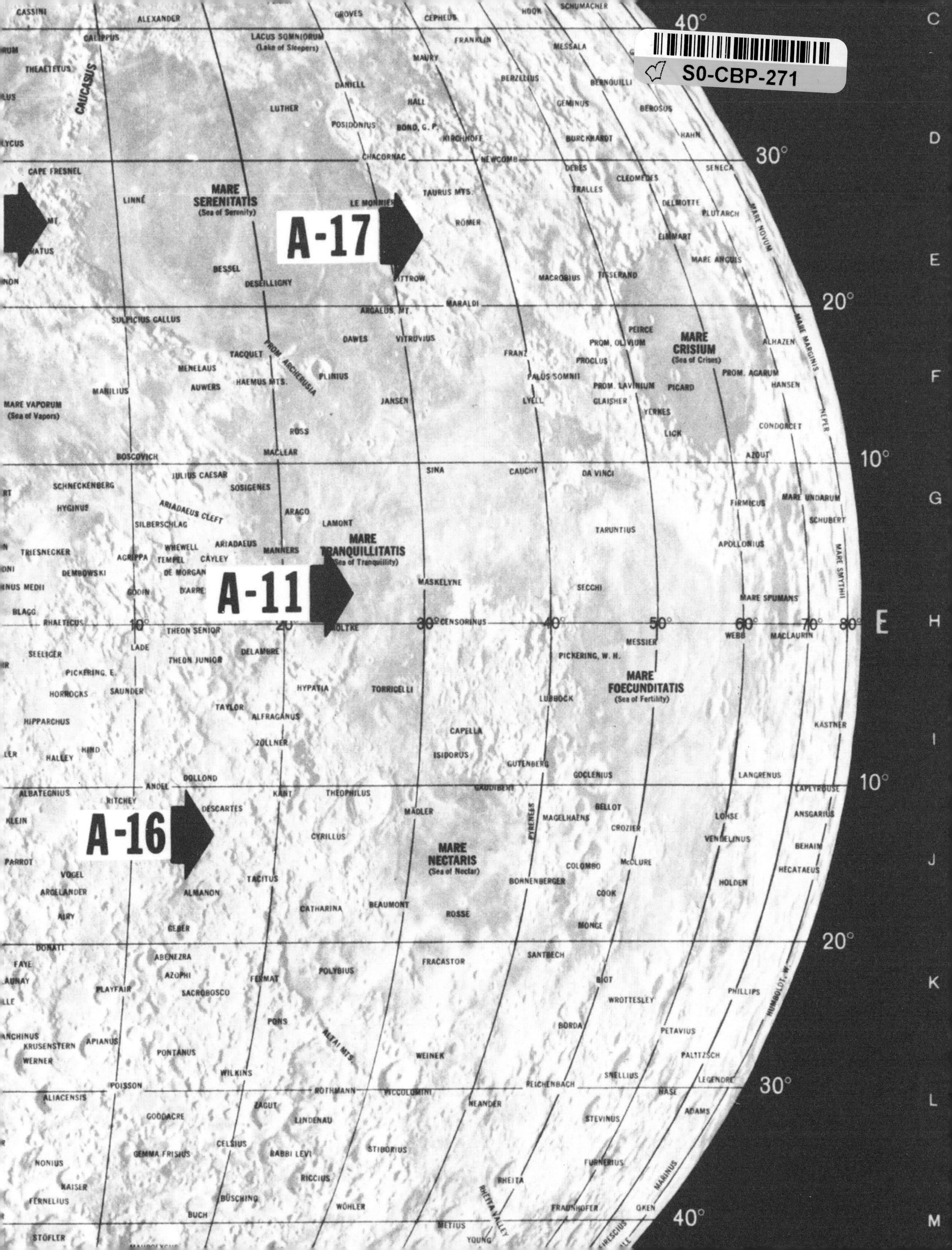

CASSINI
ALEXANDER
GROVES
CEPHEUS
HOOK
SCHUMACHER
40°
C
CALIPPUS
LACUS SOMNIORUM
(Lake of Sleepers)
FRANKLIN
MESSALA
S0-CBP-271
THEAETETUS
CAUCASUS
MAURY
DANIELL
BERZELIUS
BERNOUILLI
HALL
GEMINUS
LUTHER
BEROSUS
POSIDONIUS
BOND, G. P.
KIRCHHOFF
BURCKHARDT
HAHN
D
CHACORNAC
NEWCOMB
30°
CAPE FRESNEL
DEBES
SENECA
CLEOMEDES
TRALLES
MARE SERENITATIS
(Sea of Serenity)
LINNÉ
LE MONNIER
TAURUS MTS.
DELMOTTE
PLUTARCH
MARE NOVUM
A-17
RÖMER
EIMMART
MARE ANGUIS
E
BESSEL
DESEILLIGNY
LITTROW
MACROBIUS
TISSERAND
MARALDI
20°
ARGAEUS, MT.
SULPICIUS GALLUS
PEIRCE
DAWES
VITRUVIUS
PROM. OLIVIUM
MARE CRISIUM
(Sea of Crises)
ALHAZEN
MARE MARGINIS
TACQUET
PROM. ARCHERUSIA
FRANZ
PROCLUS
MENELAUS
PLINIUS
PALUS SOMNII
PROM. AGARUM
F
AUWERS
HAEMUS MTS.
PROM. LAVINIUM
PICARD
HANSEN
MANILIUS
JANSEN
LYELL
GLAISHER
MARE VAPORUM
(Sea of Vapors)
YERKES
KEPLER
CONDORCET
ROSS
LICK
BOSCOVICH
MACLEAR
AZOUT
10°
JULIUS CAESAR
SINA
CAUCHY
DA VINCI
SCHNECKENBERG
SOSIGENES
HYGINUS
ARIADAEUS CLEFT
ARAGO
FIRMICUS
MARE UNDARUM
G
SILBERSCHLAG
LAMONT
SCHUBERT
TARUNTIUS
MARE TRANQUILLITATIS
(Sea of Tranquillity)
TRIESNECKER
AGRIPPA
WHEWELL
ARIADAEUS
MANNERS
APOLLONIUS
TEMPEL
CAYLEY
MARE SMYTHII
DEMBOWSKI
DE MORGAN
A-11
MASKELYNE
SECCHI
GODIN
D'ARREST
MARE SPUMANS
BLAGG
RHAETICUS
10°
20°
MOLTKE
30°
CENSORINUS
40°
50°
60°
70°
80°
E
H
WEBB
MACLAURIN
THEON SENIOR
LADE
MESSIER
SEELIGER
DELAMBRE
PICKERING, W. H.
THEON JUNIOR
PICKERING, E.
MARE FOECUNDITATIS
(Sea of Fertility)
HORROCKS
SAUNDER
HYPATIA
TORRICELLI
LUBBOCK
TAYLOR
ALFRAGANUS
HIPPARCHUS
KÄSTNER
CAPELLA
ZÖLLNER
I
HALLEY
HIND
ISIDORUS
GUTENBERG
GOCLENIUS
LANGRENUS
DOLLOND
ANDEL
10°
ALBATEGNIUS
KANT
THEOPHILUS
GAUDIBERT
LAPEYROUSE
RITCHEY
DESCARTES
BELLOT
KLEIN
MÄDLER
PYRENEES
MAGELHAENS
LOHSE
ANSGARIUS
A-16
CROZIER
CYRILLUS
VENDELINUS
MARE NECTARIS
(Sea of Nectar)
BEHAIM
PARROT
J
VOGEL
COLOMBO
McCLURE
HECATAEUS
TACITUS
BOHNENBERGER
HOLDEN
ARGELANDER
ALMANON
COOK
CATHARINA
BEAUMONT
AIRY
ROSSE
GEBER
MONGE
DONATI
20°
ABENEZRA
SANTBECH
FAYE
FRACASTOR
POLYBIUS
AZOPHI
FERMAT
BIOT
K
PLAYFAIR
SACROBOSCO
PHILLIPS
HUMBOLDT, W.
WROTTESLEY
PONS
BORDA
ALTAI MTS.
PETAVIUS
KRUSENSTERN
APIANUS
PONTANUS
WEINEK
PALITZSCH
WERNER
WILKINS
SNELLIUS
LEGENDRE
POISSON
REICHENBACH
30°
ROTHMANN
PICCOLOMINI
HASE
ALIACENSIS
NEANDER
ADAMS
ZAGUT
L
GOODACRE
LINDENAU
STEVINUS
CELSIUS
GEMMA FRISIUS
RABBI LEVI
STIBORIUS
NONIUS
FURNERIUS
RICCIUS
MARINUS
KAISER
RHEITA
RHEITA VALLEY
FERNELIUS
BÜSCHING
WÖHLER
FRAUNHOFER
OKEN
BUCH
40°
METIUS
M
STÖFLER
YOUNG

Happy 10th Birthday Charlie!

Like the astronauts of Apollo 11, always dream big! Shoot for the stars because even if you don't make it, you may land on the moon :)

Love
Boppy & Grammie

# APOLLO

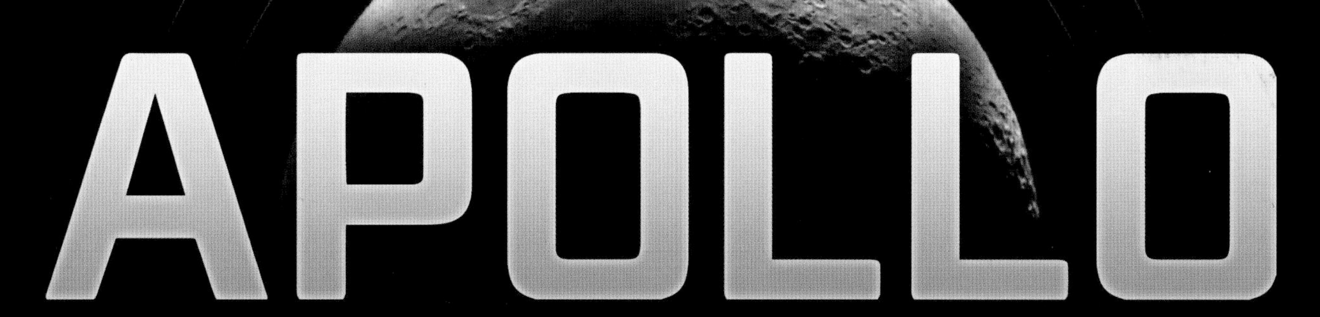

# APOLLO

## THE MISSION TO LAND A MAN ON THE MOON

AL CIMINO

CHARTWELL
BOOKS

Ed White became the first American to walk in space when he floated outside the Gemini 4 spacecraft in zero gravity for 23 minutes on June 3, 1965.

# CONTENTS

# INTRODUCTION

July 20, 2019 marks fifty years since man first set foot on the Moon in 1969. It was an extraordinary achievement. This historical spaceflight took place just twenty-five years after the first rocket reached space in the test flight of a German V-2 rocket from the Nazi research center at Peenemünde in 1944. Less than twelve years had passed since the first artificial satellite—the Soviet Union's Sputnik—had been launched in 1957. The Russian Yuri Gagarin had become the first man in space only eight years earlier.

Indeed, the youthful President John F. Kennedy had made the announcement that the Moon was the goal only eight years before. That was in May 1961, just twenty days after Alan Shepard had become the first—and then only—American astronaut to enter space on a mission that lasted just fifteen minutes. In the 1960s though, the world had recovered from the horrors of World War II and everything seemed possible. Huge technological advances had been made during the war and everyone was seized with optimism.

## THE BATTLE OVER SPACE

The space race itself was a product of the Cold War. To deflect the danger of a hot war, the communist Soviet Union and the democratic capitalist United States of America fought a symbolic battle over the conquest of space. Both sides' early launch vehicles were intercontinental ballistic missiles designed to rain down nuclear warheads on each other's cities and were based on the technology developed by the Germans in the V-2 project. The Saturn V rocket that sent Apollo 11 on its way to the Moon was designed by Wernher von Braun, who had also built the V-2 and had been captured by US troops in 1945.

The Soviets gained advantage over America with Sputnik on October 4, 1957, and the first manned flight on April 12, 1961. With the eyes of the world on the two competing superpowers, this was not a race that the US could afford to lose. So President Kennedy announced that the United States planned to send a man to the Moon, though it would still be nine months before NASA—the National Aeronautics and Space Administration—managed to put a man in orbit.

The Soviet Union were also ahead in lunar exploration. A Soviet spacecraft had already made a hard landing on the surface of the Moon and returned the first television images of the far side of the Moon, until that point, never seen by man.

Nevertheless, President Kennedy committed the United States to the mission. Against seemingly impossible odds, it was an astonishing success, which the Soviet Union simply could not match. Since Neil Armstrong first set foot on the Moon, eleven other Americans have walked on its surface. No Soviet or Russian citizen ever has.

Apollo 4 on the launch pad at the Kennedy Space Center, November 9, 1967.

PART ONE

# THE KEY TO OUR FUTURE

*WE'VE GOT TO BE FIRST. THAT'S ALL THERE IS TO IT.*

JOHN F. KENNEDY
35TH PRESIDENT OF THE UNITED STATES

CHAPTER 1

# WE CHOOSE TO GO TO THE MOON

On May 25, 1961, President John F. Kennedy concluded a speech to a joint session of Congress that the United States was planning to put a man on the Moon.

> *I believe that this nation should commit itself to achieving the goal, before this decade is out, of landing a man on the Moon and returning him safely to the Earth … No single space project in this period will be more impressive to mankind, or more important for the long-range exploration of space; and none will be so difficult or expensive to accomplish.*

He made the ideological reason for doing this perfectly clear, saying:

> *If we are to win the battle that is now going on around the world between freedom and tyranny, the dramatic achievements in space which occurred in recent weeks should have made clear to us all, as did the Sputnik in 1957, the impact of this adventure on the minds of men everywhere, who are attempting to make a determination of which road they should take.*

This was not going to be easy as Kennedy went on to explain:

> *Recognizing the head start obtained by the Soviets with their large rocket engines, which gives them many months of lead-time, and recognizing the likelihood that they will exploit this lead for some time to come in still more impressive successes, we nevertheless are required to make new efforts on our own. For while we cannot guarantee that we shall one day be first, we can guarantee that any failure to make this effort will make us last.*
>
> *We take an additional risk by making it in full view of the world, but as shown by the feat of astronaut Shepard, this very risk enhances our stature when we are successful. But this is not merely a race. Space is open to us now; and our eagerness to share its meaning is not governed by the efforts of others. We go into space because whatever mankind must undertake, free men must fully share.*

Kennedy then asked Congress for the money. Initially he asked for an extra $148 million for the space program on top of the $7 million already appropriated. Another $531 million would be needed the following year and an estimated $7 to $9 billion would be needed over the next five years. By the time the Apollo Moon-landing project had finished eleven years later, it had cost $25.4 billion, or the equivalent of $142.8 billion in 2018.

# JOHN F. KENNEDY

John Fitzgerald Kennedy (1917 - 63) was the first president to be born in the twentieth century and the first Catholic to hold that office. Entering the White House at the age of 43, he was the youngest elected president and was seen to epitomize the youthful energy and vigor of the 1960s.

He was the son of Joseph Kennedy, who had amassed a multi-million-dollar fortune in bootlegging, banking, shipbuilding, and manipulating the stock market, going on to become head of the Securities and Exchange Commission and Ambassador to Great Britain from 1938 to 1940. His son John served as his secretary for six months, giving him the material to write the bestselling book *Why England Slept*. He later published *Profiles in Courage* which won the Pulitzer Prize for Biography.

During World War II, he commanded a series of Patrol Torpedo Boats in the Pacific and was decorated for heroism after leading his crew to safety when they were marooned behind Japanese lines. He was wounded and suffered back pain for the rest of his life.

He served as congressman and senator before running for the presidency in 1960, winning by a narrow majority. His presidency was dominated by the confrontation with the Communist world, notably in the CIA-backed abortive Bay of Pigs invasion of Cuba, the Cuban missile crisis which brought the world to the edge of a nuclear war, the building of the Berlin Wall, and an escalation of fighting in Vietnam. However, he succeeded in signing a Nuclear Test-Ban Treaty with Soviet leader Nikita Khrushchev.

His assassination in Dallas in November 1963 meant that much of his work had to be completed by others—in particular the signing of comprehensive civil rights legislation into law and bringing the Apollo program to fruition.

## URGENT NATIONAL NEEDS

The speech President Kennedy made was called "Special Message to Congress on Urgent National Needs." The speech was over and above the annual "State of the Union" address because, Kennedy said: "These are extraordinary times and we face an extraordinary challenge."

In it, he mentioned the situation in Vietnam where he was already escalating US involvement in what would become the longest war in America's history. He also talked about recovery from the recession, which had taken the Dow Jones Industrial Average to its lowest point four weeks after President Kennedy was inaugurated, and social progress at home and

## Going to the Moon

*It is a most important decision that we make as a nation. But all of you have lived through the last four years and have seen the significance of space and the adventures in space, and no one can predict with certainty what the ultimate meaning will be of mastery of space.*

*I believe we should go to the Moon. But I think every citizen of this country as well as the Members of the Congress should consider the matter carefully in making their judgment, to which we have given attention over many weeks and months, because it is a heavy burden, and there is no sense in agreeing or desiring that the United States take an affirmative position in outer space, unless we are prepared to do the work and bear the burdens to make it successful. If we are not, we should decide today and this year.*

President John F. Kennedy, May 25, 1961
Special Message to Congress on Urgent National Needs

President Kennedy emerging from inside a model of the Apollo space capsule during his tour of the Manned Space Center, September 1962.

abroad. The situation in Latin America, defense and nuclear disarmament were also addressed. So the Moon landing was considered among the highest national priorities.

During the 1960s, Kennedy talked of the "missile gap," claiming that the Soviet Union was outstripping the US with Intercontinental Ballistic Missiles (ICBMs), which carried nuclear warheads. The Soviets certainly seemed to have taken the lead on the US when cosmonaut Yuri Gagarin orbited the Earth in a Vostok 1 spacecraft. Kennedy sought to raise the stakes by pouring money into the development of new ICBMs, which also served as launch vehicles in the space program, and to enhance America's battered national prestige by announcing that they planned to send a man to the Moon. It was thought that, with sufficient money and effort, the US could beat the Soviet Union to a lunar landing, while in the other prestigious project on offer—the building of a space station—they were likely to be beaten by the Soviets.

Putting a man on the Moon was not going to come cheap though, so Kennedy sought not just to appeal to Congress but to the wider public.

> *Now it is time to take longer strides—time for this nation to take a clearly leading role in space achievement, which in many ways may hold the key to our future on earth.*
>
> President John F. Kennedy, May 25, 1961

## JFK'S SPEECH AT RICE STADIUM

NASA's Manned Spacecraft Center—later the Lyndon B. Johnson Space Center—was to be established in Houston, Texas, on land donated through Rice University. It was at the university's football stadium on September 12, 1962, that President Kennedy decided to make another speech during a whistle-stop tour of NASA's new facilities, this time exclusively about the advance of science and the space program. Why, he asked rhetorically, had they chosen the Moon as their goal? He told a cheering crowd of 40,000 on a hot day:

*We choose to go to the Moon. We choose to go to the Moon in this decade and do the other things, not because they are easy, but because they are hard, because that goal will serve to organize and measure the best of our energies and skills, because that challenge is one that we are willing to accept, one we are unwilling to postpone, and one which we intend to win, and the others, too.*

*It is for these reasons that I regard the decision last year to shift our efforts in space from low to high gear as among the most important decisions that will be made during my incumbency in the office of the Presidency.*

*In the last twenty-four hours we have seen facilities now being created for the greatest and most complex exploration in man's history. We have felt the ground shake and the air shattered by the testing of a Saturn C-1 booster rocket, many times as powerful as the Atlas which launched John Glenn, generating power equivalent to ten-thousand automobiles with their accelerators on the floor. We have seen the site where the F-1 rocket engines, each one as powerful as all eight engines of the Saturn combined, will be clustered together to make the advanced Saturn missile, assembled in a new building to be built at Cape Canaveral as tall as a forty-eight-story structure, as wide as a city block, and as long as two lengths of this field.*

*Within these last nineteen months at least forty-five satellites have circled the earth. Some forty of them were "Made in the United States of America" and they were far more sophisticated and supplied far more knowledge to the people of the world than those of the Soviet Union.*

*The Mariner spacecraft now on its way to Venus is the most intricate instrument in the history of space science. The accuracy of that shot is comparable to firing a missile from Cape Canaveral and dropping it in this stadium between the forty-yard lines.*

The bright spiral galaxy IC 342 is the sort of far-distant body of stars to which humans dream of traveling in the future. It is ten million light-years away from Earth.

# CHAPTER 2

# DREAMS OF EARTH AND SKY

Humans have long dreamed of traveling to the Moon. In *A True Story*, the second-century Greek writer Lucian described being carried there by a whirlwind to find that its inhabitants were at war with those who lived on the Sun.

Some ancient peoples believed the Moon was a bowl of fire, while others thought it was a mirror that reflected Earth's lands and seas, but ancient Greek philosophers knew the Moon was a sphere orbiting the Earth whose moonlight reflected sunlight.

The Greeks also believed the dark areas of the Moon were seas while the bright regions were land, which influenced the current Latin names for those places—"maria" for seas and "terrae" for land.

In 1608, the German astronomer Johannes Kepler wrote the novel *Somnium* (*The Dream*) in which Duracotus, the son of an Icelandic witch, is transported to the Moon via a lunar bridge. From there, he observes the movements of the Earth that show Kepler's theories on planetary motion are true.

The Italian astronomer Galileo Galilei was the first to use a telescope to make scientific observations of the Moon. In 1609 he described a rough, mountainous surface that was quite different from the popular beliefs of his day that the Moon was smooth.

## THE SCIENCE OF FLIGHT

Kepler's contemporary, Francis Godwin, Bishop of Llandaff and Hereford, wrote *The Man in the Moone*, where the narrator devises a flying machine that harnesses the power of giant swans to carry him there.

In 1640, scientist and theologian Dr. John Wilkins drew up plans for what he called a flying chariot powered by clockwork and springs, a set of flapping wings coated with feathers, and a few gunpowder boosters to help send it on its way.

However, he believed that the gravitational and magnetic pull of the Earth only extended for twenty miles. Once you had broken through that you could float out into space. What's more, space travelers would not have to take any food with them as he believed that people on Earth only needed to eat because gravity was constantly pulling food through the body, emptying the stomach.

By the 1660s, British scientists Robert Boyle and Robert Hooke demonstrated that there was a vacuum stretching between the Earth and the Moon, making space travel problematic. Wilkins later came to realize that magnetism and gravity were not the same thing. So there was not going to be what Wilkins had described as a "sphere of magnetic virtue" twenty miles around the Earth.

The French novelist Cyrano de Bergerac famously traveled using rockets in *The Other World: Comical History of the States and Empires of the Moon* published in 1657. In *The Consolidator* (1705) Daniel Defoe flew his narrator to the Moon in a machine.

## STRANGER THAN FICTION

Jules Verne used a huge gun to fire men to the Moon in *From Earth to the Moon* in 1865. The fictional gun was sited on a hill in Florida not far from the present-day Kennedy Space Center. H.G. Wells got in on the act in 1901 with *The First Men on the Moon* after the discovery of Cavorite, a substance that negates gravity.

In 1928, Dr. Dolittle was carried to the Moon by a huge moth. Written in the 1940s, Arthur C. Clarke's first novel *Prelude to Space* narrated in a realistic manner the first manned voyage to the Moon by the rocket Prometheus—a complex two-stage spacecraft powered by a nuclear reactor. Stronger on technology than character, it was explicitly written to encourage the belief that landing on the Moon was possible.

Tintin, Snowy, Captain Haddock, and Professor Calculus all set foot on our sister planet in Hergé's *Explorers on the Moon* in 1954. By 1966, sci-fi writer Robert A. Heinlein had set a penal colony on the Moon in his novel *The Moon is a Harsh Mistress*.

## BY MEANS OF ROCKET DEVICES

The Russian scientist Konstantin Tsiolkovsky was the first to consider the practicality of space exploration. In 1903, he published "Explorations of Outer Space by Means of Rocket Devices" in Russia's scientific review.

The article introduced the Tsiolkovsky equation used to calculate the horizontal speed needed to maintain an orbit around the Earth. This, he said, could be achieved by a multi-staged rocket fueled by liquid hydrogen and liquid oxygen. Tsiolkovsky developed these ideas in subsequent articles, but they had little impact outside Russia.

Inspired by Tsiolkovsky's work, French aircraft designer Robert Esnault-Pelterie worked out the energy required to reach the Moon and nearby planets. He went on to experiment with ballistic missiles, but failed to spark much interest in rocketry in France.

---

> *So far this is a thankless, risky and extremely difficult task. It will strain our resources and talents to the limit, and it will entail great losses too ... one cannot compare spaceflight to aeronautics. The latter is like a toy compared to the former.*
>
> Konstantin Tsiolkovsky
> Russian space scientist, 1903

---

Meanwhile in the US, Robert H. Goddard was working along similar lines. In 1919, he published *A Method of Reaching Extreme Altitudes*. His claim that rockets could be used to send objects as far as the Moon was ridiculed in the press, including *The New York Times*, who finally recanted on July 17, 1969, after Apollo 11 had taken off on its way to the Moon.

Even when he raised his own funding and began experimental work, Goddard was mocked. After one test-flight in 1929, a local Worcester newspaper carried the mocking headline "Moon rocket misses target by 238,799½ miles."

Goddard then moved to Roswell, New Mexico, where he could carry on his experiments unhindered. By 1937, he was reaching altitudes of 8,000 to 9,000 meters (2,500 to 2,700 feet, over 1.5 miles).

In Germany, Austrian physicist and engineer Hermann Oberth published *Ways to Spaceflight* in 1929 and launched his first rocket, using a liquid-propellant motor, on May 7, 1931. He was assisted by a young student named Wernher von Braun. In 1941, Oberth joined the rocket development center at Peenemünde, where he worked under von Braun. Later Oberth worked on rocketry for the US Army.

# KONSTANTIN TSIOLKOVSKY

Born in Izhevskoye, 120 miles (193 km) south-east of Moscow, Konstantin Eduardovich Tsiolkovsky (1857 – 1935) lost his hearing when he was 9 years old as a result of scarlet fever. Unable to attend school, he studied mathematics at home and, while still a teenager, began to speculate on space travel.

At age 16, he went to Moscow where he studied chemistry, mathematics, mechanics, and astronomy, and became interested in the science of flight. He became a teacher in Borovsk, about 60 miles (96 km) from Moscow, where he continued his studies on his own. His paper on the kinetic theory of gases earned him membership of the Russian Physio-Chemical Society.

Moving to Kaluga, 100 miles (160 km) south-west of Moscow, he turned his attention to aeronautics. Working on the construction of an all-metal dirigible, he built Russia's first wind tunnel, which earned him a grant from the Academy of Sciences. In 1895, he published *Dreams of Earth and Sky*. In 1903, he published his most important work, "Exploration of Outer Space by Means of Rocket Devices."

Despite their support, the Academy of Sciences ignored the results of his work. After the Russian Revolution in 1917, the Soviet state continued to support his research and he played a significant role in the development of astronautics. Elected to the Academy of Sciences, he was granted a pension for life for his contribution.

# ROBERT H. GODDARD

Born in Worcester, Massachusetts, the imagination of Robert Hutchings Goddard (1882 – 1945) was fired after reading H.G. Wells' novel *The War of the Worlds* which was serialized in the *Boston Post* in 1898. With ideas from the book fresh in his head, he climbed a cherry tree and imagined the possibility of ascending to Mars.

"I was a very different boy when I descended the tree from when I ascended," he said. The date was October 19, 1899, which he remembered as Anniversary Day.

He studied physics at Clark University in Worcester, earning a PhD and becoming a full professor there in 1920. Meanwhile, his notebooks were filled with designs for multi-staged rockets, with stages being discarded once the propellant was exhausted.

In 1914, he was granted the first of his 214 patents, all but six related to rockets and spaceflight. In 1916, the Smithsonian Institution gave him a grant of $5,000 to study rockets as a means to carry instruments into the upper atmosphere. When the US joined World War I in 1917, he was given $20,000 to work on rocket-propelled weapons. His work ended with the armistice, but his research was used in the development of the bazooka in World War II.

He was the first to develop a rocket motor using liquid hydrogen and liquid oxygen. A rocket using liquid fuel lifted off from his Aunt Effie's Farm in Auburn, Massachusetts, on March 16, 1926. Further funding was raised by aviation pioneer Charles Lindbergh, and the Guggenheim Fund for the Promotion of Aeronautics funded a small shop and experimental flights in Roswell, New Mexico.

By 1935, his liquid-fueled rockets were breaking the sound barrier and reaching an altitude of one mile (1.6 km). During World War II, his Roswell shop was closed down and he worked on the development of a jet booster for seaplane takeoff in Annapolis, Maryland. Years after his death in 1945, the US government made a settlement of $1 million for his patents.

# WERNHER VON BRAUN

Wernher von Braun (1912 – 77) became interested in space and astronomy after his mother gave him a telescope when he was confirmed in the Lutheran Church. His fascination with rocketry came from the speed records then being set up by rocket-powered cars. At age 13, he got a copy of *The Rocket into Interplanetary Space* by German rocket pioneer Hermann Oberth. This inspired him to apply himself to physics and mathematics, and he went on to work on Oberth's first rocket in 1931.

The following year he began work on solid-fuel rockets for the Ordnance Department of the German Army. While the Treaty of Versailles ending World War I restricted German weaponry, it did not mention rockets. By 1934, von Braun's team were achieving altitudes of 1.5 miles (2.4 km). He joined the Nazi Party in 1937.

When the Kummersdorf Army Proving Grounds near Berlin became too small, they moved to Peenemünde, where von Braun became technical director. There he worked on rocket-powered aircraft and the A-4, a long-range ballistic missile named the Vengeance Weapon 2, or V-2, by the Propaganda Ministry.

After von Braun and his team surrendered to US troops in 1945, they were put to work at the US Army Ordnance test site at White Sands, New Mexico. At Huntsville, Alabama, they developed ballistic missiles. He was granted US citizenship in 1955 and his group launched the first US satellite, Explorer I, in 1958.

They were transferred to NASA when it was formed in 1958 and developed large-scale launch vehicles, including the Saturn V used in the Apollo program.

On June 16, 1977, Wernher von Braun died of pancreatic cancer in Alexandria, Virginia, at age 65. His gravestone quotes Psalm 19:1: "The heavens declare the glory of God; and the firmament sheweth his handywork."

## THE WEAPONS AT PEENEMÜNDE

The first rockets capable of traveling into space were developed in Nazi Germany at Peenemünde on the Baltic coast. The weapons research center was established there in 1937. The technical director was Wernher von Braun.

The V-1 flying bomb and the V-2 ballistic missile were developed there, produced using slave labor. They were fired against Britain, France, and Belgium in the dying days of World War II. As the war came to a close, von Braun was working on the V-3, which would have brought the US within range. Work was moved to underground facilities in the Austrian Alps to escape Allied bombing.

Bumper Wac liftoff at Cape Canaveral, Florida, 1950. Bumper Wac was a two-stage experimental rocket which employed a V-2 as a first stage and a Wac Corporal upper stage. It was designed by the German rocket team to provide data for upper atmospheric research.

*I imagined how wonderful it would be to make a device which had the possibility of ascending to Mars.*

Robert H. Goddard when he was a boy, on the afternoon of October 19, 1899

At the end of World War II, the US, Britain, and the Soviet Union scrambled to capture as many V-2s and rocket engineers as they could. The US managed to snatch parts for a hundred V-2s from what would become the Soviet Zone of Germany before the Red Army arrived. Von Braun and his team also surrendered to the US. They were taken to the White Sands Proving Grounds in New Mexico.

## DREAMING ABOUT ROCKETS

The V-2 parts were studied and some twenty-five V-2s were assembled. These were used as sounding rockets to carry instruments into the upper atmosphere. In 1946, one produced the first photograph taken in outer space. The following year, a gyroscope in another failed and it crashed over the border in Mexico.

Despite occasional failures, one V-2 reached an altitude of 114 miles (183 km). As parts ran out, General Electric made new components and over seventy-five of these V-2 sounding rockets were fired.

Von Braun, though, was keen to get on with the exploration of outer space. To do that he would need the support of the American people. He told a colleague:

> *We can dream about rockets and the Moon until Hell freezes over. Unless the people understand it and the man who pays the bill is behind it, no dice.*

Von Braun set out the case for space in the book *The Mars Project*, which was rejected by eighteen American publishers. It was eventually published by the University of Illinois Press in 1953, but it did not find a wide audience. However, during the process, he met Cornelius Ryan, author of *The Longest Day*, at a conference of space medicine in San Antonio. Ryan was also an associate editor of *Collier's*, one of America's most popular magazines.

Over the next two years, *Collier's* published eight articles on space exploration. Von Braun made a major contribution and became a household name. He appeared on television, notably with Walt Disney collaborating on a series of three educational films.

## THE MISSILE RACE

In August 1949, the Soviet Union tested an atomic bomb in the desert of Kazakhstan. Four years later, in August 1953, the Soviets announced they had developed a hydrogen bomb. In the event of war, these would have to be delivered by aircraft. However, unlike the US, they did not have a vast fleet of long-range bombers. Any attack on the US would be a suicide mission. What was needed was the rapid development of an intercontinental ballistic missile (ICBM). By February 1956, the Soviets' chief designer Sergei Korolev had developed the R-7, the world's first ICBM, which could reach US targets in less than thirty minutes.

Meanwhile US efforts were in disarray. The Air Force had given up on its Atlas missile and was working on the Titan. The Navy's Vanguard program was a catastrophic failure, with eight of the first launches failing. The first Thor ICBM failed immediately after liftoff, crashed back onto the pad and exploded. The second went off course and had to be destroyed. The third exploded four minutes before it was due to launch. The fourth broke up after ninety-two seconds. And three of the Jupiters von Braun built for the Army zigzagged off course.

Nevertheless, on September 20, 1956, the two-stage Jupiter did manage a record-breaking flight of 3,400 miles (5,470 km). However, the Defense Secretary decided that the remaining Jupiters be handed over to the Air Force to be deployed in Italy and Turkey, as protection against the Soviet threat. Meanwhile the Army and von Braun were restricted to battlefield weapons with a range of 200 miles or less.

Von Braun was in despair, saying publicly: "I expect we will have to pass Russian customs when we finally reach the Moon."

## The Cold War

There had been antagonism between the US and the Soviet Union since its beginnings in 1917 when Communist revolutionaries took over the Russian Empire and executed the Tsar.

During World War II, the US and the Soviet Union found themselves on the same side fighting Adolf Hitler and Nazi Germany in Europe. As the war drew to its conclusion, the Soviet Red Army occupied much of Eastern Europe and imposed Communist rule there.

The US and its Western Allies feared that the Soviet Union would try to take over Western Europe too and formed the North Atlantic Treaty Organization (NATO), for mutual defense. The Soviet Union and the Communist nations in Eastern Europe responded with the Warsaw Pact.

After the Soviet Union developed an atomic bomb in 1949, there was a genuine fear that there would be a nuclear war between the rivals. Huge stockpiles of weapons were built. Smaller proxy wars between the Communist and Free Worlds were fought in Korea, Vietnam, and elsewhere.

The cost of weaponry eventually crippled the Soviet Union. The Communist regimes in Eastern Europe collapsed. The Soviet Union itself collapsed in 1991. Fifteen independent nations were formed instead, including Russia which briefly enjoyed a period of democracy.

Tanks carrying missiles during the annual military display in Red Square, Moscow, helped fuel the Cold War between the Soviet Union and the USA.

# CHAPTER 3

# WONDERS OF SCIENCE

At 10:26 p.m. on October 4, 1957—on the hundredth anniversary of Konstantin Tsiolkovsky's birth—a two-stage R-7 was launched from Baikonur Cosmodrome in Kazakhstan. The launch was at night so that any failure would be hidden from the eyes of the American spy planes that regularly flew overhead. But if it was a success, it would be a wonder of science and the whole world would see it.

The 100 feet (30 m) of the rocket's casing had been covered with prisms to catch the light of the Moon. At an altitude of 139 miles (223 km) it attained the orbital velocity of 18,000 mph (29,000 kph), and the R-7 jettisoned a metal ball approximately the size of a basketball with four trailing antennas. This was Спутник-1, or Sputnik 1—the world's first artificial satellite.

It gleamed like a star in the sky and could clearly be seen as it passed over America. It also sent out a radio signal and its "meep ... meep ... meep" could be heard by anyone with a ham radio set. The implication was clear.

"Thanks to Comrade Korolev and his associates, we now have a rocket that could carry a nuclear warhead," said Soviet First Secretary Nikita Khrushchev, adding: "His invention also has many peacetime uses."

---

*Launching of the satellite was a tremendous victory for science. It was a more tremendous victory for Soviet propaganda to be able to trumpet to the world the Russians were the first to break through the frontiers of space.*

United Press, October 5, 1957

---

## A RACE FOR SURVIVAL

Sputnik came as a complete surprise. While Korolev could read translated newspaper reports of US missile tests, von Braun did not even know of Korolev's existence. The military and the CIA knew, but the distribution of the intelligence was restricted. President Dwight D. Eisenhower was unimpressed.

"The Russians, under a dictatorial society, where they have some of the finest scientists in the world, who have for many years been working on it, apparently from what they say they have put one small ball in the air," he said. Indeed, the feeling in the White House was that if the Soviet Union could send Sputnik over America, the US could send spy satellites over Soviet territory—if they ever launched one.

Others were more fearful. Senate Majority Leader and future US President Lyndon Johnson fumed that the Russians could one day be "dropping bombs on us from space like kids dropping rocks onto cars from freeway overpasses." Physicist Edward Teller, the "father of the hydrogen bomb," described the event as "a greater defeat for our country than Pearl Harbor."

US Congresswoman Clare Boothe Luce called it "an intercontinental outer-space raspberry to a decade of American pretensions that the American way of life was a gilt-edged guarantee of our material superiority." The press also sought to stir up public anxiety. The *Chicago Daily News* said:

> *The day is not far distant when they [the Soviets] could deliver a death-dealing warhead onto a predetermined target almost anywhere on the earth's surface.*

Even *The New York Times* warned that unless immediate measures were taken to put the United States in the lead again, the race would be "not so much a race for arms or even prestige, but a race for survival."

The anxiety was international. British Prime Minister Harold Macmillan told the House of Commons:

> *I say without hesitation and without excuse that this is a turning point in history. Never has the threat of Soviet Communism been so great, or the need for countries to organize themselves against it.*

*Time* magazine condemned Eisenhower's sanguine response with a piece headlined "A Crisis of Leadership" and the Dow fell ten percent.

## BIGGER PAYLOADS

The red lights eventually went on in the White House the following month when Sputnik 2 carried the dog Laika to an altitude of 1,031 miles (1,659 km), proving that the Soviets could launch bigger payloads, such as a hydrogen bomb. In fact, there was no immediate danger as the Soviets had not yet solved the problem of reentry, though America did not know that.

Until then the CIA had insisted that there was no "missile gap," based on their overflights of U-2 spy planes. Now it was predicted that the Soviet Union would have a hundred ICBMs by 1959, 500 by 1960, and would land on the Moon in 1965. The President's Science Advisory Committee immediately recommended that the government spend $25 billion on fallout shelters. That would be $222 billion at 2019 prices. Leaflets recommending shelters with walls one-foot-thick were sent out to America's suburbs on the assumption that inhabitants of the inner cities could do nothing to save themselves.

*If we lose repeatedly to the Russians as we have lost with the Earth satellite, the accumulated damage will be tremendous. We should accordingly plan, ourselves, to achieve the next big breakthrough first, a manned satellite, or getting to the Moon.*

Arthur Larson
Dwight D. Eisenhower's presidential speechwriter

Khrushchev turned the screw by telling the official Soviet newspaper *Pravda*:

> *The fact that the Soviet Union was the first to launch an artificial earth satellite, which within a month was followed by another, says a lot. If necessary, tomorrow we can launch ten, twenty satellites. All that is required for this is to replace the warhead of an intercontinental ballistic rocket with the necessary instruments. There is a satellite for you.*

And he said that the Soviet Union would be turning out R-7s "like sausages."

## SEIZING THE MOMENT

Clearly, the US had to get a satellite into orbit as soon as possible to save face. The US Navy stepped up to the plate. A small satellite, 6.4 inches (16.3 cm) in diameter and weighing just 3 pounds (1.36 kg), was mounted on top of a three-stage Vanguard rocket that stood 75 feet (23 m) high.

On December 6, 1957, the public and the world's press were invited to America's Atlantic Missile Range at Cape Canaveral in Florida to witness the launch. However, about two seconds after liftoff, when the rocket had risen about 3 feet (1 meter) the booster

# SERGEI KOROLEV

Sergei Pavlovich Korolev (1906 – 66) studied aeronautical engineering under the celebrated designers Andrey Nikolayevich Tupolev and Nikolay Yegorovich Zhukovsky before becoming interested in rocketry. He was a founder member of the Moscow Group for the Study of Reactive Motion, which launched the Soviet Union's first liquid-fueled rocket in 1933, going on to work on the cruise missiles and a rocket-powered glider at the Jet Propulsion Research Institute in Leningrad, now St. Petersburg.

In June 1938, he was arrested for slowing the work of the institute. He was tortured and imprisoned, while others were executed, putting Soviet rocketry far behind the advances being made in Nazi Germany. After months in the gulag where he lost most of his teeth, he was sent to a special prison that was essentially a slave-labor camp for engineers and scientists. During World War II he designed bombers under fellow prisoner Tupolev.

After the war, he was set to work test-firing captured V-2s and modifying them to increase their range. He then headed the Soviet Union's ballistic missile program that led to the development of the first ICBM. He also headed the Soviet program to produce launch vehicles and both unmanned and manned spacecraft.

During his lifetime, he was only known as the "chief designer." His identity was only revealed after he was dead. He died of abdominal cancer in 1966, and is buried in the walls of the Kremlin on Red Square, Moscow.

# SPUTNIK

The Soviet Union inaugurated the space age by launching Sputnik 1 on October 4, 1957. It was a pressurized sphere of aluminum alloy, 23 inches (58 cm) in diameter and weighing 184 pounds (83.6 kg). Its scientific objectives were to test the method of placing an artificial satellite into Earth orbit, provide information on the density of the atmosphere by calculating its lifetime in orbit, test radio and optical methods of orbital tracking, determine the effects of radio wave propagation through the atmosphere, and check principles of pressurization used on the satellites. Its significance though was propaganda—it demonstrated that the Soviet Union was ahead of the US in both the missile and space race.

The word Sputnik originally meant "fellow traveler," but it has now become synonymous with satellite in modern Russia. Its elliptical orbit took it to a high point of 584 miles (940 km) above the surface of the Earth and a low point of 143 miles (230 km). The orbit took ninety-six minutes.

Sputnik 1 transmitted for twenty-one days until its batteries were exhausted. It remained in orbit for ninety-six days, then it fell back and burned up in the Earth's atmosphere. Sputnik 2 was launched on November 3, 1957, and carried the dog Laika, or "Barker," the first living creature to be sent into space. She died of overheating after only a few hours, but there was no way she could have been returned safely to Earth. Sputnik 2 reentered the Earth's atmosphere after 162 days.

Eight more Sputniks were sent up carrying a variety of animals to test life-support systems prior to manned flights. The program also gathered data on radiation, temperature, pressure, and magnetic fields, as well as testing reentry procedures.

*Oh little Sputnik, flying high*
*With made-in-Moscow beep,*
*You tell the world it's a Commie sky*
*And Uncle Sam's asleep.*

Mennen Williams
Governor of Michigan

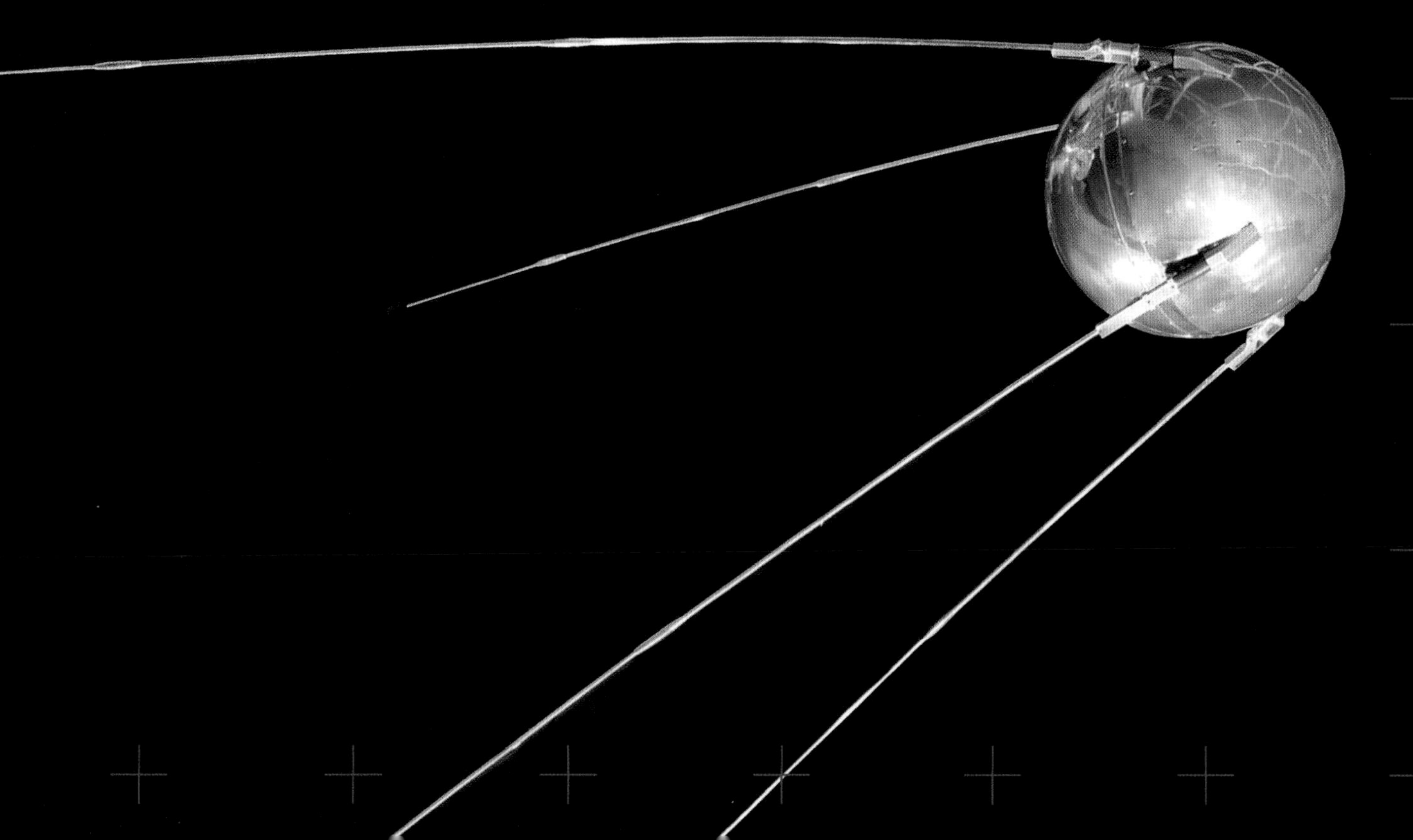

lost thrust and began to settle back on the launch pad. As it settled, the fuel tanks ruptured and exploded, destroying the rocket and severely damaging the launch pad.

The satellite was thrown clear and landed in the brush a short distance away with its transmitters still sending out a beacon signal, causing one reporter to say: "Why doesn't somebody go out there, find it, and kill it?" As it was, the satellite was damaged beyond repair and could not be reused.

Wernher von Braun seized the moment. He told the new Secretary of Defense Neil McElroy:

> *When you get back to Washington you'll find that all hell has broken loose. I wish you would keep one thought in mind through all the noise and confusion: We can fire a satellite into orbit sixty days from the moment you give us the green light.*

## LAUNCHING EXPLORER 1

On January 31, 1958, the 18-pound (8 kg) Explorer 1 stood on top of a Juno launch vehicle which was adapted from a Jupiter-C sounding rocket. This was a joint venture between von Braun's group in Huntsville, Alabama, and the Jet Propulsion Laboratory in Pasadena, California. Although it was clearly an Army operation, due to the humiliation of the military the previous month, the launch was dressed up as a civilian enterprise.

A Jupiter-C rocket launches Explorer 1, the United States' first successful satellite, January 31, 1958.

With the shells of the upper stages spinning at 550 rpm to act as a gyroscopic stabilizer, the first stage burned for 157 seconds. When the craft reached an altitude of 60 miles (96 km), explosive bolts blew, allowing it to fall away into the Atlantic. The second stage burned for 247 seconds, the third for 6.5 seconds. When they fell away, the satellite's own booster ignited. Explorer 1 reached the required escape velocity of 18,000 mph (29,000 kph) and went into orbit.

Explorer 1 was a more sophisticated craft than Sputnik. It carried instruments to detect the Van Allen radiation belts, returning data for four months until the batteries ran out. It remained in orbit until 1970 and was followed by ninety similar craft. However, it was small and the Juno launch vehicle was not powerful enough to carry a nuclear warhead. Khrushchev returned Eisenhower's "little ball" slight, calling Explorer 1 a "grapefruit."

Von Braun was seen as the hero of the hour. His autobiography was run as a serial in *The American Weekly*. His life story was then made into the 1960's movie *I Aimed at the Stars*, inspiring the stand-up comedian Mort Sahl's quip: "Wernher von Braun aimed at the stars, but often hit London." It was only fifteen years after World War II ended, and von Braun was still a controversial figure.

## FROM NACA TO NASA

Despite the successful launch of Explorer 1, America's space program was still in disarray with the Army, Navy, and Air Force in competition for resources. But there was another player, the National Advisory Committee for Aeronautics (NACA).

NACA was a federal agency set up in 1915 to undertake aeronautical research. Orville Wright sat on the board in the 1920s when it provided wind tunnels and other aircraft test facilities. Its research facilities proved vital during World War II. NACA ran the tests and development of the Bell X-1 which broke the sound barrier.

By 1958, NACA had 8,000 employees and $300 million worth of facilities when James Killian, chairman of the President's Science Advisory Committee, recommended that it should be turned into the National Aeronautics and Space Administration—NASA.

# X-PLANES

NACA, later NASA, developed a series of rocket-engined X-planes. The first was the Bell X-1 which was the first plane to break the sound barrier in the hands of the famous test pilot Chuck Yeager. Variants increased the top speed to Mach 2.21.

Bell's X-2 reached 3.196 Mach, or 2,094 mph (3,370 kph). The Douglas X-3 Stiletto barely reached Mach 1, but led to the development of the Lockheed F-104 Starfighter that operated at Mach 2 and was used by the USAF during the Vietnam War.

Developments continued up to the X-15. Plans were conceived to make this the first space rocket, to be launched on top of a SM-64 Navaho missile. This was canceled with the development of the Mercury project. Among the test pilots who flew it was Neil Armstrong.

In July and August 1963, pilot Joseph A. Walker exceeded 62.1 miles (100 km) in altitude, joining NASA astronauts and Soviet cosmonauts in crossing the official line marking the beginning of outer space, though the USAF awarded astronaut wings to anyone achieving an altitude of 50 miles (80 km).

Boeing also began the development of the X-20 Dyna-Soar spaceplane to put manned military spacecraft into orbit. It was decades ahead of its time, and might have given the Air Force a manned spaceplane more than two decades before the space shuttle. But the Dyna-Soar also fell victim to the Mercury spacecraft program and was canceled by Secretary of Defense Robert McNamara before it ever had a chance to fly.

Famous for being the first man on the Moon, before becoming an astronaut, Neil Armstrong (pictured) risked life and limb in a variety of experimental vehicles. He flew seven missions in the X-15 as a NASA test pilot, reaching a top speed of Mach 5.74 or 4,044 mph (7,087 kph) and a peak altitude of 39.2 miles (63 km).

# CHAPTER 4

# WHAT GOES UP MUST COME DOWN

On July 29, 1958, President Eisenhower signed the National Aeronautics and Space Act, establishing NASA which began operations on October 1. NASA absorbed NACA intact with its annual budget of $100 million, along with three major research laboratories—Langley Aeronautical Laboratory, Ames Aeronautical Laboratory at Edwards Air Force Base in California, and Lewis Flight Propulsion Laboratory. It also included two small test facilities—the Wallops Island launch facility and the Navy's Research Laboratory, which became the Goddard Space Flight Center, both in Maryland.

Pasadena's Jet Propulsion Laboratory and the Army Ballistic Missile Agency at Huntsville that had launched Explorer 1 were added, along with all units dealing with the non-military exploration of space. Meanwhile the Advance Research Projects Agency was created to develop the military applications of space technology.

## SHOOTING AT THE MOON

The same year that NASA was created, the Soviet Union began its program of unmanned Moon shots. Initially the idea was simply to hit the Moon, rather than orbit it. The first three failed due to faults with the booster rocket. The fourth—then dubbed Luna 1—managed to escape Earth's gravity, but missed the Moon and went into orbit around the Sun.

The fifth failed due to a problem with the flight control, but the seventh—Luna 2—became the first man-made object to hit the Moon on September 13, 1959. Then on October 7, 1959, Luna 3 transmitted back the first pictures of the other side of the Moon. During that time, NASA had tried twenty-eight times to get a satellite into orbit, with only eight successes. Two of them were lunar flybys, Pioneer 3 launched on December 6, 1958, and Pioneer 4 on March 3, 1959.

Luna 9 made the first soft landing on the Moon on February 3, 1966, after a series of failures. NASA followed, making a soft landing with Surveyor 1 on May 30, 1966. Although the Soviets were still clearly ahead in the space race, NASA was catching up fast.

# THE PIONEER PROGRAM

The Pioneer probes were unmanned spacecraft designed for planetary exploration. Launched on August 17, 1958, Pioneer 0 was a lunar orbiter, but was destroyed when the first stage of the Thor booster blew up shortly after liftoff. As it was such a failure it was not called Pioneer 1, though the actual Pioneer 1 was almost as disastrous, when a partial failure of the third stage meant it missed the Moon.

The third stage failed to ignite on Pioneer 2 and it fell back into the Earth's atmosphere. The next Pioneer was destroyed in an explosion on the launch pad. The one following that failed soon after launch. Pioneer 3, launched on December 6, 1958, reached an altitude of 63,600 miles (102,360 km), then fell back to Earth. Pioneer 4 flew by the Moon, passing within 36,650 miles (58,983 km), but not close enough to trigger its instruments.

Pioneer 5, launched on March 11, 1960, was successful and went on to explore the interplanetary space between Earth and Venus. Pioneers 6 through 9 went into orbit around the Sun, making observations. Launched on December 16, 1965, Pioneer 6 was still working in the year 2000.

Pioneer 10, launched on March 12, 1972, was sent to photograph Jupiter, making its closest approach on December 4, 1973. Contact was lost on January 23, 2003, when it was 7.5 billion miles (12 billion km) from Earth.

Launched on April 6, 1973, Pioneer 11 was also sent to investigate Jupiter and Saturn. Contact was lost on November 24, 1995. Two more Pioneer probes—12 and 13—were sent to Venus.

Pioneer 10 blasts off in 1972 on the first successfully completed mission to the planet Jupiter. The first spacecraft to traverse the asteroid belt, Pioneer 10 transmitted about 500 photographic images of Jupiter from a range of 16 million miles (25 million km). After thirty-one years in space, communications were eventually lost with Pioneer 10 in 2003.

# THE SURVEYOR PROGRAM

The unmanned Surveyor spacecraft were sent to the Moon to gather data on the surface prior to the Apollo landings. Before Luna 9, four months ahead of the landing of Surveyor 1, it was not known how deep the lunar dust was. If it was too deep, no astronaut could land. The Surveyors were launched on an Atlas-Centaur rocket that put it directly onto a translunar flight path. As it approached the Moon it fired retro-rockets that slowed it from 5,840 mph (9,400 kph) to 248 mph (400 kph) before steerable vernier motors took over.

The Surveyors took around two-and-a-half-days to reach the Moon. Of the seven missions, five were successful. Surveyor 3 made an unintentional take-off from the surface when its vernier motors did not shut down. Surveyor 6 was the first spacecraft to make an intentional liftoff.

## SUCCESS TURNS TO FAILURE

The Soviets scored another victory in August 1960 when they sent two dogs, Strelka and Belka, into space. They spent a day orbiting the Earth before landing safely. They were accompanied by rats, mice, and insects, along with plants and sheets of human skin to test the effects of radiation. All the passengers survived. Meanwhile NASA suffered more failures, including the explosion of its first Mercury-Atlas rocket 58.5 seconds after take-off.

Khrushchev was so pleased with the success of the Soviet program that he gave out models of Russian spacecraft. Eisenhower was given a model of a Luna module, while the Secretary-General of the United Nations Dag Hammarskjöld was presented with one of the Martian probe Mars 1 in 1960. Later, in 1962, President Kennedy received one of Strelka's puppies.

Then the Soviets had a series of unfortunate disasters. After three rocket failures, the prototype of the Soviet's latest R-16 rocket blew up on the launch pad, killing over a hundred engineers, technicians, and military personnel on October 24, 1960.

After that on December 2, 1960, Sputnik 6, carrying two more dogs Pchyolka and Mushka, failed on reentry when its retro-rockets failed to shut off. The explosives that ensured the capsule did not fall into foreign hands were detonated, and all the experimental animals onboard died.

The failure of the R-16 delayed the ICBM program, so Khrushchev sought to place his existing IRBMs (intermediate-range ballistic missiles) closer to targets in the US. This move would later spark the Cuban missile crisis. Meanwhile, on May 1, 1960, a U-2 spy plane was shot down over Russia and the pilot, Gary Powers, was captured, so President Eisenhower decided that work on the development of military reconnaissance satellites be stepped up.

## ASTRONOMICAL COSTS

On August 18, 1960, the spy satellite Corona 14 filmed more of the Soviet Union than all the U-2 spy-plane missions put together. It was a huge success. The US could now get an accurate count of how many ICBMs the Soviet Union had—far fewer than had been estimated.

President Eisenhower was a military man. He could understand the necessity of expenditure to obtain such vital intelligence. But NASA's plan to put a man in

Mercury-Redstone 1 (MR-1) was intended to be an unmanned suborbital spaceflight. But on November 21, 1960, the launch failed when the engine shut down immediately after liftoff. The rocket only rose about 4 inches (10 cm) before settling back onto the pad.

# SPY SATELLITES

The US Corona spy satellite was launched on the Thor-Agena rocket. In a polar orbit, it would fly over the Soviet Union filming anything bigger than the size of a truck. After it reentered the atmosphere, it would be slowed by parachutes and captured by a USAF plane, or recovered in the Pacific by the US Navy. Publicly Corona would be known as the Discoverer scientific research satellite, though those with security clearance called it Keyhole.

The first launch was scheduled for January 21, 1959. The Agena stage accidentally separated on the launch pad, aborting the mission. This was followed by a further twelve failures. Only on August 18, 1960, the very day U-2 pilot Gary Powers was standing in the dock of a court in Moscow, did Corona 14 successfully return with film that showed 1.5 million square miles of Soviet territory, including sixty-four airfields and twenty-six surface-to-air missile launchers, along with the launch pads at Baikonur and Plesetsk.

From May 1960, the US launched Midas and Vela with infrared sensors and other detectors watching for the exhaust of a launching missile or contraventions of the 1963 Test-Ban treaty.

The Soviets had their own spy satellites—the Zenit, publicly designated as Kosmos—which was similar in design to the Vostok manned spacecraft. The first flight on December 11, 1961, had to be aborted. More failures followed, but on July 28, 1962, Kosmos 7 was launched successfully and returned with photographic surveillance of the entire United States.

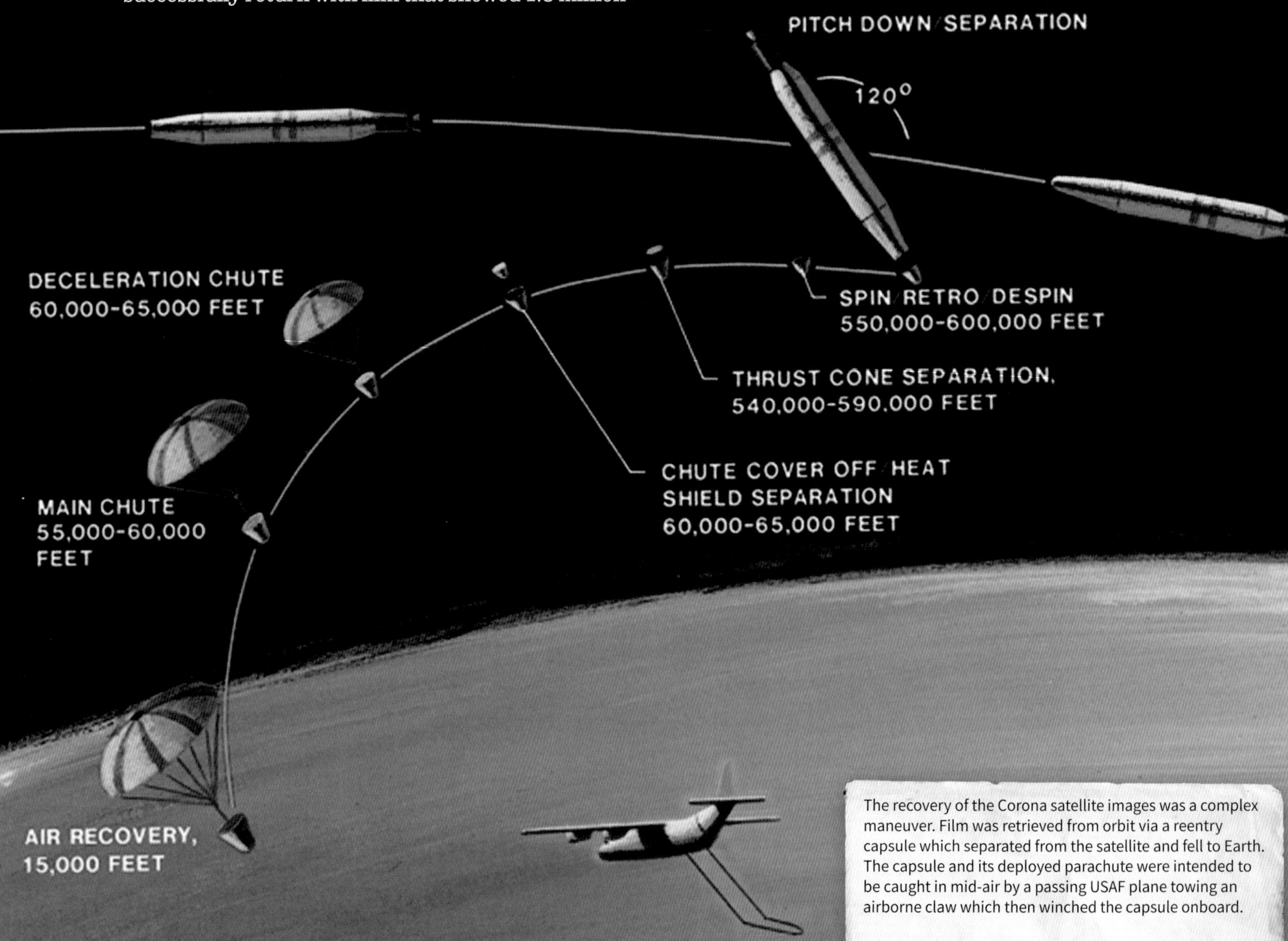


The recovery of the Corona satellite images was a complex maneuver. Film was retrieved from orbit via a reentry capsule which separated from the satellite and fell to Earth. The capsule and its deployed parachute were intended to be caught in mid-air by a passing USAF plane towing an airborne claw which then winched the capsule onboard.

orbit and then send men to the Moon passed him by. He said he did not care whether man ever reached the Moon and he even talked about terminating Project Mercury, the program to put a man into space.

When he asked for the reasons to undertake such ambitious and expensive programs, Robert Seamans of NASA compared sending men to the Moon to Queen Isabella of Spain financing Christopher Columbus's voyages to the new world.

But the costs were literally astronomical. Project Mercury to put a man in orbit was estimated to cost $350 million, an orbit of the Moon $8 billion, and a manned lunar landing another $26 to $38 billion—$290 billion to $390 billion at 2018 prices. Eisenhower said that he was not about to hock the family jewels to send men to the Moon.

In a draft of President Eisenhower's last budget in 1960, he said that there was no scientific or defense need for a man-in-space program beyond Mercury. However, his scientific advisor Dr. George Kistiakowsky suggested that it would be unwise to make such a blunt statement. Instead in his final budget message, Eisenhower said: "Further test and experimentation will be necessary to establish if there are any valid scientific reasons for extending manned spaceflight beyond the Mercury program."

Eisenhower may have been lukewarm about going into space, but his successor John F. Kennedy would be thrilled by the prospect.

> *If we let scientists explore the Moon, then before you know it, they'll want funds to explore the planets.*
>
> White House aide, December 20, 1960

## THE MISSILE GAP

When Eisenhower left office in January 1960, the US had 160 Atlas ICBMs, plus nearly a hundred IRBMs deployed in Europe, while the Soviets has just four R-7s. During the year four US reconnaissance satellites over-flew the Soviet Union, reducing the estimate of its missiles from 120 to a maximum of fifty and probably as few as fourteen. However, the Soviets' lead in the space race had a powerful grip on the minds of the American people.

Senator Kennedy, who was running for president that year against Eisenhower's Vice President Richard Nixon, exploited this as a perceived "missile gap." He maintained that the US falling behind was the fault of the previous Republican administration, implicating Nixon. This was reflected in the Soviet Union's lead in the space race. On September 7, 1960, Kennedy said:

> *The people of the world respect achievement. For most of the twentieth century they admired American science and American education, which was second to none. But they are not at all certain about which way the future lies. The first vehicle in outer space was called Sputnik, not Vanguard. The first country to place its national emblem on the moon was the Soviet Union, not the United States. The first canine passengers in space who safely returned were named Strelka and Belka, not Rover or Fido, or even Checkers.*

Checkers was famously the name of Richard Nixon's dog who he had introduced to the America people in a TV address in 1952 when he had been accused of misusing campaign funds when running for vice president.

Later that month the magazine *Missiles and Rockets* addressed an open letter to both candidates on the issue. Kennedy responded immediately, saying:

> *We are in a strategic space race with the Russians, and we are losing. If a man orbits Earth this year, his name will be Ivan. If the Soviets control space they can control the Earth, as in past centuries the nation that controlled the seas has dominated the continents. We cannot afford to run second in this vital race. To ensure peace and freedom we must be first. Space is our great New Frontier.*

Kennedy had adopted the slogan "New Frontier" in his speech accepting the presidential nomination at the Democratic Convention in Los Angeles earlier that year. He explained that the New Frontier was:

> *. . . the frontier of the 1960s, the frontier of unknown opportunities and perils, the frontier of unfilled hopes and unfilled threats . . . Beyond that frontier are uncharted areas of science and space, unsolved problems of peace and war, unconquered problems of ignorance and prejudice, unanswered questions of poverty and surplus.*

Kennedy, of course, went on to win the election.

PART TWO

# THE NEW FRONTIER

*THE MOON IS THE FIRST MILESTONE ON THE ROAD TO THE STARS.*

ARTHUR C. CLARKE
SCIENCE FICTION WRITER AND FUTURIST

# CHAPTER 5

# THE SPACE RACE

Once in office, President Kennedy did not accelerate the space race at first, but rather he sought to use the exploration of space as a means of détente. In his inaugural address on January 20, 1961, he said:

> *Let both sides seek to invoke the wonders of science instead of its terrors. Together let us explore the stars, conquer the deserts, eradicate disease, tap the ocean depths and encourage the arts and commerce.*

In his State of the Union address ten days later, Kennedy went further, saying:

> *I now invite all nations—including the Soviet Union—to join with us in developing a weather prediction program, in a new communications satellite program and in preparation for probing the distant planets of Mars and Venus, probes which may someday unlock the deepest secrets of the universe.*

The cooperation was needed because:

> *Today this country is ahead in the science and technology of space, while the Soviet Union is ahead in the capacity to lift large vehicles into orbit.*

However, Kennedy was more concerned about the Pentagon's need for missiles and the CIA's need for reconnaissance satellites than manned missions. Nevertheless he appointed Vice President Lyndon Johnson chairman of the National Aeronautics and Space Council which oversaw the federal government's space programs. He had been chairman of the Senate Aeronautical and Space Sciences Committee and, as Senate majority leader, he had used his influence to pass the 1958 National Aeronautics and Space Act, setting up NASA.

Next he needed a new head of NASA. He picked Washington insider James Webb who would handle relations with the White House, Congress, and the American people. Hugh Dryden would oversee science and Robert Seamans was general manager. The manned spaceflight division would be headed by D. Brainerd Holmes, who would oversee von Braun in Huntsville, the Manned Spacecraft Center in Houston, and the Launch Operations Center at Cape Canaveral, later renamed the Kennedy Space Center.

# JAMES WEBB

Born in Tally Ho, North Carolina, James Edwin Webb (1906 – 92) served as a pilot in the US Marine Corps before studying law at George Washington University and being admitted to the bar of the District of Columbia. In 1936, he went to work for the Sperry Gyroscope Corporation, which made aeronautical instruments.

Ten years later he went to work at the treasury, then became director of the Bureau of the Budget and, in 1949, undersecretary of state, also serving as an informal foreign-policy advisor to Senator Robert S. Kerr. In 1960, when Lyndon Johnson was elected vice president, Kerr replaced him as chairman of the Senate Aeronautical and Space Sciences Committee and Johnson recommended Webb be appointed head of NASA, then considered a political backwater. More than a dozen candidates had turned the job down.

Webb was in place when President Kennedy decided NASA should put a man on the Moon. He defended NASA in Washington after a fire onboard Apollo 1 killed three astronauts and amid growing criticism of its expenditure during the Vietnam war. Webb resigned in 1968 after Johnson said he would not seek reelection, fearing that his association with Johnson and Kennedy would damage the agency under a Republican administration. He was succeeded by his non-partisan deputy Thomas Paine.

Johnson awarded Webb the Presidential Medal of Freedom and he went on to give a series of lectures at Columbia University, which were published as *Space Age Management: The Large-Scale Approach* in 1969. He stayed on in Washington, serving on several advisory boards and as regent of the Smithsonian Institution. He was buried in Arlington National Cemetery and was honored after his death in 1992 when the successor to the Hubble Telescope was named the James Webb Space Telescope.

## ROCKETRY AND SPACECRAFT

NASA was now dedicated to manned spaceflight. Its launch site would be alongside the Air Force base at Cape Canaveral. Even though it was plagued with bad weather and corrosive salt in the air, it was easily accessible by water, so rockets and other large assemblies could be delivered by barge.

The Manned Spacecraft Center was established at Houston, Texas, so that it was easy to reach from the spaceport in Florida, major manufacturers in California, and MIT in Massachusetts who supplied many of the scientists. Texas was also the home of Lyndon Johnson, one of the foremost advocates of the space program, and Congressman Albert Thomas, chairman of the House Appropriations Committee who opposed some of Kennedy's key legislation. The President got his bills. Thomas was also associated with an oil company who were persuaded to give a thousand acres of land 22 miles (35 km) south-east of Houston to his alma mater Rice University who then donated them to NASA.

At first glance the site did not look too promising. Hurricane Carla had just devastated the area. When the Army Corps of Engineers moved in, all they found were a few shacks, ramshackle windmills, and a lonesome cowboy skinning a wolf.

The ex-NACA staff from Langley, Virginia, had to move to Houston, where they were joined by others from Florida. The engineers drafted in had particular problems. Most had worked on aircraft flight tests and knew nothing about spacecraft or rocketry. "We had to virtually invent or adapt every tool we used," said flight director Gene Kranz.

*In August of 1961, I found myself in Washington with an illustrious title, an organization of somewhat conflicting interests, and the challenge of establishing a program to send a man to the Moon and bring him safely home. It was a formidable assignment.*

D. Brainerd Holmes
Head of Manned Spaceflight Division, NASA

## Hidden Figures

The story of three of the African-American women who worked at NASA during the Space Race—Katherine Johnson, Mary Jackson, and Dorothy Vaughan—was told in the movie *Hidden Figures*. NASA research mathematician Sandra Jansen remembers the era of the human computers:

*Computing at that time consisted of row after row of women ... who sat and did line after line of calculations on desktop calculators. I was a math major, so I could take the equations and translate them into the various sheets they needed to do their job. They didn't have to do the math; all they had to do was follow the instructions ... The girls who worked there were called computers.*

The human computers, left to right: Katherine Johnson, Mary Jackson, and Dorothy Vaughan.

## THE FIRST ASTRONAUT

NASA's first astronaut was a 3-year-old chimpanzee named Ham who went into space on January 31, 1961. Using banana pellets and electric shocks, he had been taught to perform simple tasks. However, when he was blasted off from Cape Canaveral in a Mercury capsule on top of a Redstone rocket the system of rewards and punishments malfunctioned and he was getting a shock when he had done everything perfectly.

The flight path was over one degree too high, subjecting Ham to 15g (gravitational force)—3g more than planned—as the speed rose to 5,857 mph (9,426 kph) when it should have been 4,400 mph (7,081 kph). Then the cabin depressurized and the temperature dropped, but in his small space suit Ham felt no ill-effects. The spacecraft overshot and splashed down 60 miles (96 km) from the nearest rescue ship. By the time helicopters arrived, the capsule was taking on water and sinking. Nevertheless Ham was rescued and when he arrived on the deck of USS *Donner* he readily accepted his reward of an apple and half an orange.

Ham lived out the remaining years of his life in the National Zoo in Washington and a zoo in North Carolina. It was clear that NASA could not risk sending a man into space without further test flights carrying chimps.

Though Ham was cooperative, others were not. One named Enos managed to splatter a visiting Congressman's white shirt and tie with dung.

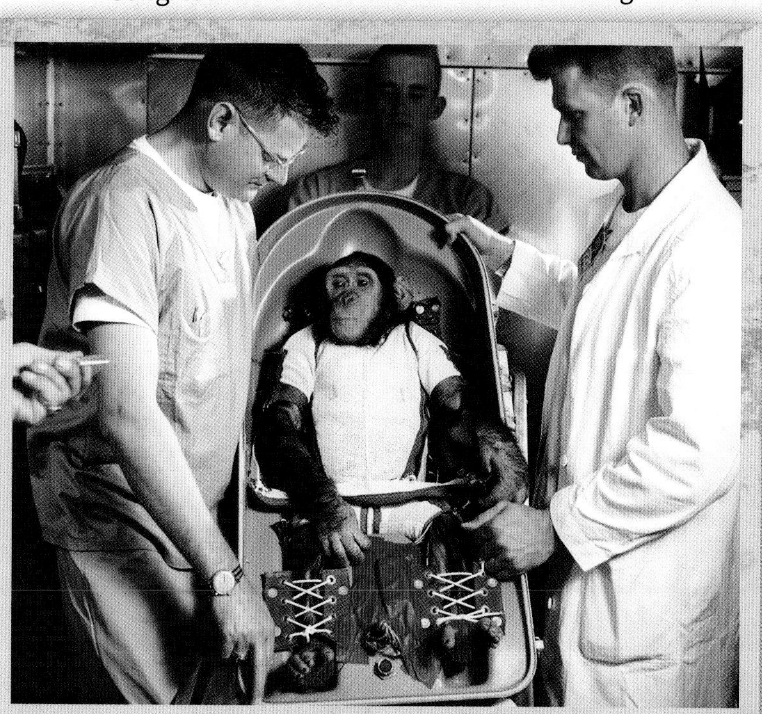

Ham, the chimpanzee, is strapped into a biopack couch in preparation for his suborbital spaceflight.

## Pigs in Space

While designing the Mercury capsule a large number of pigs were used in drop tests as the insides of a pig are much like those of a human being. According to engineer Günter Wendt: "What they would do is make a couch for pigs, then hoist them up on a rather tall facility, and drop them and see what happened to the crushable material. And some of the tests went very well. Some of them didn't go too well. But then whatever was left over went to the orphanage."

McDonnell in St. Louis used so many pigs that they had a pigpen in the factory. Early in the Mercury program, there were problems, because the engineers knew nothing about pigs. Aerospace technician Alan Kehlet recalled:

> *We were going to launch a pig, and we put him in the cradle and started monitoring him, and the pig died. One of our secretaries was a farm girl, and she said, "If you'd asked me before you had the pig in there, I would have told you that you never put a pig on his back, that the belly fat on there will suffocate the pig." And that's exactly what happened. So we went from pigs to monkeys, and the monkey was kind of interesting, too, because he would get an electric shock if he didn't perform his duties rightly.*

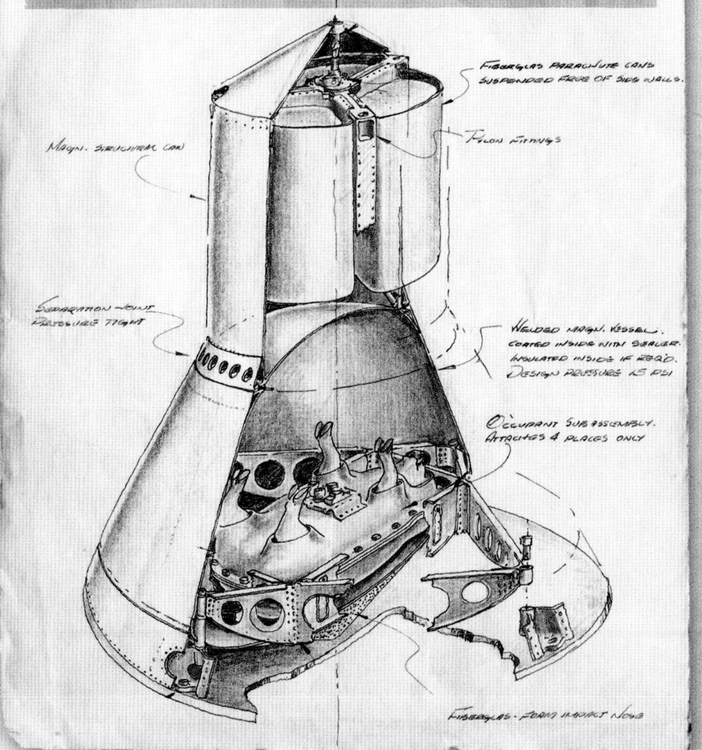


The initial proposed arrangement for a pig couch in a Mercury capsule.

## TALK OF NATIONAL EXTINCTION

In March 1961, Webb asked Kennedy for a steep hike in NASA's budget, warning that, otherwise, the Soviets would remain ahead in the space race for five to ten years. He also took the opportunity to propose a lunar landing by the end of the decade. But NASA recorded in the early 1960s that only a modest increase was approved. Günter Wendt recalled that they were making between twenty and thirty launches a week, and sixty percent were failures. Wendt said, "At one time, the seven astronauts were here watching an Atlas launch, and that thing blew, too. Shepard said, 'I hope they fix that problem before they launch us.' "

The Soviets did not seem to be troubled by such problems. On the morning of April 12, 1961, Air Force Colonel Yuri Gagarin was blasted into space onboard Vostok 1 and orbited the Earth once at a maximum altitude of 187 miles (301 km) and landed safely 1 hour 48 minutes later.

"Now let other countries try to catch us," said Gagarin.

Again there was consternation in America. *Life* magazine compared the US to the decaying Roman Empire being overrun by the Visigoths. Lyndon Johnson pointed out that the Romans created their empire by building roads. The British Empire had flourished because it had ships. The US became powerful early in the twentieth century because it had planes. Now the Soviets had a foothold in space.

*The Earth was gay with a lavish palette of colors. It had a pale blue halo around it. Then this band gradually darkened, becoming turquoise, blue, violet then coal-black.*

Yuri Gagarin
The First Man in Space

In Congress there was talk of "national extinction." Congressman James Fulton asked Hugh Dryden whether the Soviets could produce red dust and turn the Moon red. Dryden conceded they could. Kennedy came under pressure at a press conference. He said he was tired of trailing behind the Soviets and that he hoped America was "going to go into other areas where we can be first and which will bring perhaps more long-range benefits to mankind."

## PLAYING CATCH-UP

Two days later at a meeting in the White House, Kennedy asked: "Is there any place we can catch them? What can we do? Can we put a man on the Moon before them?"

Dryden replied that if the US were to make putting a man on the Moon the equivalent of the Manhattan Project to make an atomic bomb, at around $40 billion, they had a fifty-fifty chance of succeeding. White House Counsel and one of Kennedy's speechwriters Ted Sorensen, who was at that meeting, said Kennedy "immediately sensed that the possibility of putting a man on the Moon could galvanize public support for the exploration of space as one of the great human adventures of the twentieth century."

In the week following that meeting, the Kennedy administration suffered another setback. A CIA-backed invasion of Cuba by anti-communists at the Bay of Pigs failed disastrously. This was on top of reversals in Laos and the Congo. On April 20, Kennedy sent a memo to Lyndon Johnson, which read:

1. Do we have a chance of beating the Soviets by putting a laboratory in space, or by a trip around the Moon, or by a rocket to land on the Moon, or by a rocket to go to the Moon and back with a man? Is there any other space program which promises dramatic results in which we could win?
2. How much additional would it cost?
3. Are we working twenty-four hours a day on existing programs? If not, why not? If not, will you make recommendations to me as to how work can be speeded up.
4. In building large boosters should we put our emphasis on nuclear, chemical, or liquid fuel, or a combination of these three?
5. Are we making maximum effort? Are we achieving necessary results?

## THE CHANCES OF SUCCESS

Johnson met with NASA executives who told him that the Soviets were likely to be the first to build a space station, but the US could put the first man on the Moon if a determined effort was made. This would mean increasing the decade's budget of just $12 million to between $22 and $34 billion.

Such an investment would also help the aerospace industry which was suffering a downturn as the missile race was slowing. Indeed President Eisenhower had originally started the space program to help the aerospace industry who were then suffering a post-war slump in demand.

As head of the George C. Marshall Space Flight Center in Huntsville, Wernher von Braun also received a copy of Kennedy's memo. His answers were:

a.) We do not have a good chance of beating the Soviets to a "manned laboratory in space."

b.) We have a sporting chance of beating the Soviets to a soft landing of a radio transmitter station on the Moon.

c.) We have a sporting chance of sending a three-man crew around the Moon ahead of the Soviets.

d.) We have an excellent chance of beating the Soviets to a first landing of a crew on the Moon (including return capability, of course).

## WORDS OF WARNING

Although he was not asked his opinion, von Braun added some of his own words of warning as a caveat, telling Johnson that:

> *... in the space race we are competing with a determined opponent whose peacetime economy is on a war time footing ... I do not believe that we can win this race unless we take at least some measures which thus far have been considered acceptable only in times of a national emergency.*

Plainly Kennedy ignored this warning as he made his announcement to Congress the following month. Johnson was also undaunted: "Would you rather have us be a second-rate nation, or should we spend a little money."

While it was more than a little money that had to be found, the US had already been spending vast amounts to fund fighting World War II, the Manhattan Project, and in the aftermath, pay for the Marshall Plan that had distributed $13 billion work of economic aid.

Top: Russian cosmonaut Yuri Gagarin became the first man in space when he orbited the Earth aboard Vostok 1 on April 12, 1961. Bottom: Vostok 1 blasting off from the Baikonur Cosmodrome.

Top: Alan B. Shepard Jr. became the first American to travel into space in 1961.
Bottom: *Freedom 7* lifts off on May 5, 1961, carrying Alan Shepard on his 15-minute suborbital flight.

The CIA also reported that, while the Soviets were likely to orbit the Moon by 1966, they would not be able to land on it until 1970. But James Webb grew anxious that concentrating on the race to the Moon would divert NASA from scientific research, as well as launching communications and weather satellites, and sending unmanned probes to survey the Moon.

## FIRST AMERICAN IN SPACE

On May 5, 1961, astronaut Alan B. Shepard Jr. became the first American to fly into space, buckling himself into his Mercury capsule *Freedom 7* after being reminded by Chuck Yeager, the test pilot who had taken the Bell X-1 through the sound barrier, not to forget to "brush the monkey shit off the seat." Although the flight only lasted 15 minutes 22 seconds, forty-five million Americans watched him and it caught the public's imagination.

Afterward, he went to visit President Kennedy with a delegation of executives from NASA. Kennedy listened intently as Shepard described the flight. Toward the end of the conversation, Kennedy asked: "What are we doing next? What are our plans?"

The men from NASA then talked about going to the Moon and Kennedy said: "I want a briefing." Shepard described Kennedy as a real "space cadet."

The director of the Manned Spacecraft Center Bob Gilruth said: "The president was impressed by the world's reaction to the Shepard flight and wanted to know more about what we are going to do."

He and his deputy George Low outlined the rest of the Mercury project and briefed Kennedy on NASA's recent study of sending a man in orbit around the Moon.

"Why aren't you considering landing men on the Moon?" Kennedy wanted to know. "If we're going to beat the USSR, don't we need to do something more than just flying around the Moon?"

Gilruth pointed out that landing on the Moon was an order-of-magnitude harder than orbiting around it. Kennedy immediately asked him what they would need.

"Sufficient time, presidential support, and a congressional mandate," said Gilruth.

"How much time?" asked Kennedy.

"Ten years," Gilruth replied.

## JOURNEY TO MARS

The following day, James Webb and Robert Seamans visited Defense Secretary Robert McNamara at the Pentagon to prepare a memo on what would become the Apollo project. Webb explained that NASA was planning to land on the Moon by the end of the 1960s. McNamara insisted that this was short-sighted and they should proceed directly to Mars.

Webb and Seamans were horrified. The reality was that NASA at that time did not even have the capability to contemplate a landing on the Moon, far less trying to land on Mars. Nevertheless, McNamara was convinced that Apollo would help out the US aerospace industry which was reeling from cuts by the Department of Defense. Kennedy, too, was persuaded by this argument.

Lyndon Johnson also argued that a non-military, non-commercial, and non-scientific mission would enhance the nation's prestige. Space had captured the world's imagination and dramatic achievements would demonstrate the superiority of their system over communism.

## WE'VE GOT TO BE FIRST

Soon after, Kennedy called Webb, Dryden, and Gilruth to the White House. Kennedy told them: "All over the world we're judged by how well we do in space. Therefore, we've got to be first. That's all there is to it."

Gilruth figured that Kennedy was going to tell them to cut out any further suborbital flights—NASA planned six of them—and move straight on to orbital missions, using the Atlas rocket. They were shocked when Kennedy said: "I want you to start on the Moon program. I'm going to ask Congress for the money. I'm going to tell them that you're going to put a man on the Moon by 1970."

## The Berlin Wall

At the end of World War II, the victors divided Germany into four zones of occupation. The Soviet Union occupied the east, while the US, Britain, and France occupied the west. The German capital, Berlin, while a hundred miles inside the Soviet zone, was also divided by the three powers.

In 1949, Western powers united their zones to form the Federal Republic of Germany, or West Germany, including their unified zones of Berlin, which became West Berlin. In response, the Soviet zone became the German Democratic Republic, or East Germany. In 1946, Winston Churchill had made a speech in Fulton, Missouri, saying: "From Stettin in the Baltic to Trieste in the Adriatic, an iron curtain has descended across the Continent."

The only place that the iron curtain was left open was in Berlin. Hundreds of thousands of people took advantage of this to seek asylum in the West. On August 13, 1961, the border between East and West Berlin was closed and the construction of a wall dividing the city began.

The Berlin Wall was torn down in 1989, heralding the collapse of communism in eastern Europe, the reunification of Germany, and the demise of the Soviet Union.

The Berlin Wall was built across the middle of Potsdamer Platz which had once been a busy city square and traffic intersection before being destroyed by bombing during World War II.

*I could hardly believe my ears. I was literally aghast at the size of the project being undertaken. At the time, there was no detailed plans and no studies in depth on how the landing could be done.*

Bob Gilruth
Director of the Manned Spacecraft Center
on hearing Kennedy's speech, May 25, 1961

That was exactly what he did in his speech to the joint session of Congress on May 25. NASA was given $1.7 billion for the following year, though it was clear that it could have any amount of money it wanted and the space program went into a period of "budgetless financing."

The decision was not without its critics. Eisenhower called it "just nuts." Hauled in front of the Senate Appropriations Committee, Hugh Dryden was asked what purpose would be served by landing on the Moon. He replied: "It certainly does not make sense to me." Even President Kennedy had his misgivings. He would have preferred to spend the money on a project that would have done some good on Earth, but space was the only place the US could challenge the Soviet Union without the risk of war. That is why it would become the nation's priority.

However, both before his speech to Congress and after it, Kennedy made approaches to the Soviet Union to abandon the space race and cooperate in exploring the universe. But the Soviets were not interested. Khrushchev said that they had to pursue military disarmament before they could begin joint missions in space.

Meanwhile the Cold War got colder. On August 6, 1961, the USSR put a second cosmonaut into space, who orbited the Earth seventeen times. A week later, work on the Berlin Wall began. While the Kennedys in Hyannis Port and the von Brauns in Huntsville began building radiation-proof nuclear fallout shelters, space again became the only arena where the competition between the US and the USSR could be fought out.

## The Cuban Missile Crisis

In May 1960, Soviet Premier Nikita Khrushchev promised to defend the newly installed regime of Fidel Castro and his communist supporters. In July 1962, the Soviet Union began shipping medium- and intermediate-range ballistic missiles to Cuba. On October 14, 1962, U-2 over-flights saw launch sites being prepared. President Kennedy decided to blockade the island. As the Soviet ships carrying the missiles approached, the world hovered on the brink of nuclear war for a week. At the last moment, the Soviet ships turned away. The Soviet Union agreed to remove nuclear missiles already installed on Cuba, while the US agreed not to make any further attempts to invade the island, and to remove its own nuclear missiles from bases in Turkey and Italy.

Aerial photograph of medium-range ballistic missile (MRBM) launch site number 3, San Cristobal, Cuba.

# THE MERCURY SEVEN

The Mercury Seven were the seven astronauts who flew in the Mercury project announced by NASA on April 9, 1959. Pictured below are front row, left to right: Walter M. Schirra, Jr., Donald K. "Deke" Slayton, John H. Glenn, Jr., and M. Scott Carpenter; back row, Alan B. Shepard, Jr., Virgil I. "Gus" Grissom, and L. Gordon Cooper, Jr.

After appearing on the cover of *Life* magazine, the Mercury Seven were treated like movie stars. Their homes were invaded. They were inundated with fan mail and their phone numbers had to be changed regularly. But they were paid just $5,500 to $8,000, with a $2,000 housing allowance and $1,500 in flight pay. The technicians got more as the astronauts were on military pay scales.

As they could not get life coverage from insurance companies, Houston congressman Albert Thomas tried to set up a special insurance scheme for the astronauts, but this was deemed to be unfair to all the men drafted to fight in Vietnam who could not get life insurance either. *Life* magazine provided $50,000 life insurance in exchange for exclusive rights to their stories, and they gave them an allowance for clothing as the astronauts no longer wore uniforms and had little civilian clothing.

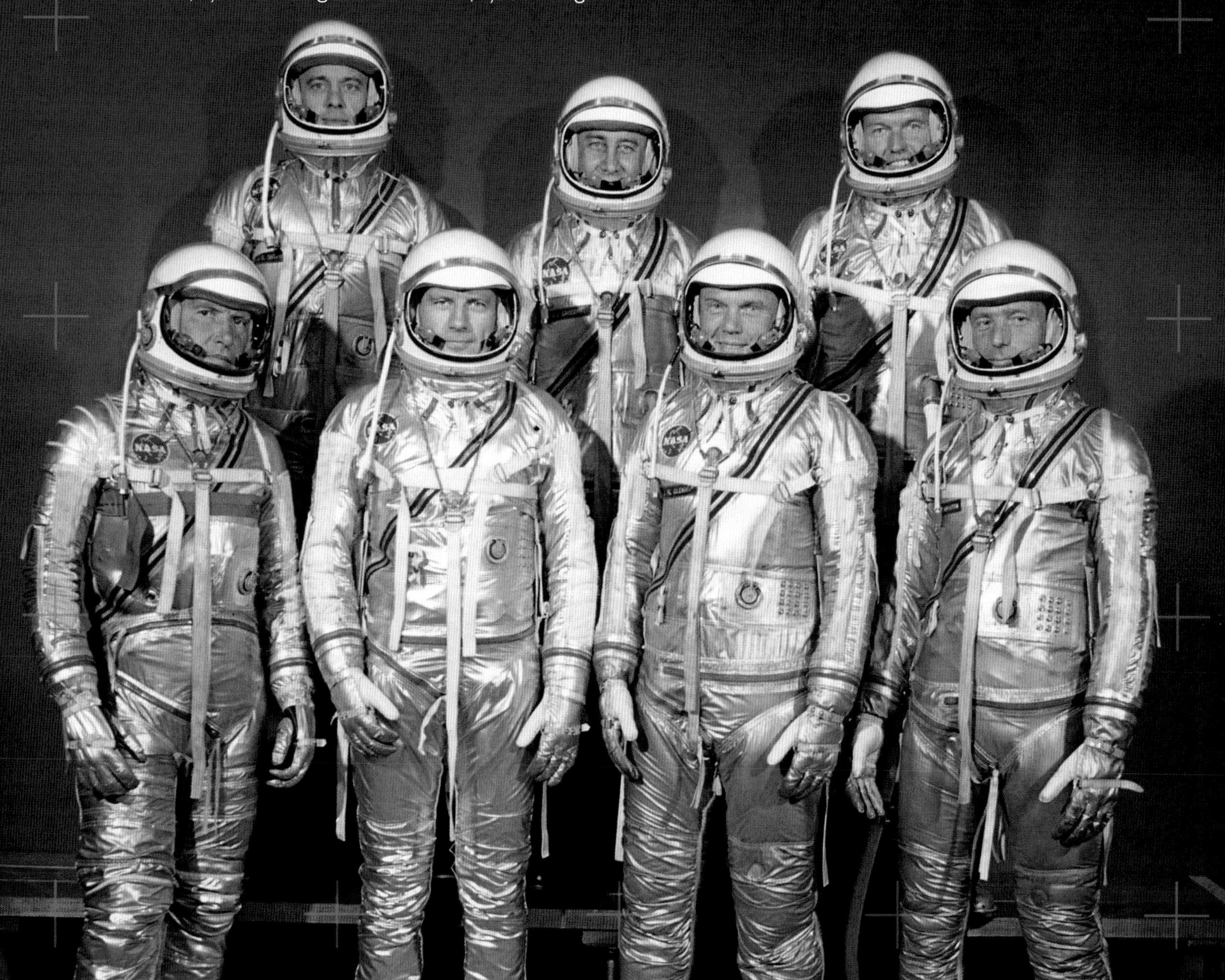

CHAPTER 6

# MEETING THE CHALLENGE

President Kennedy had told the world that NASA was going to put a man on the Moon when it had just fifteen minutes experience of manned spaceflight. In fact, the Apollo program had already started before Alan Shepard went into space. In July 1960, NASA began a feasibility study into using spacecraft carrying crews of more than one man to build a space station, fly around the Moon, orbit it, or make a manned landing there.

Contracts worth $250,000 were given to General Electric, and aircraft makers Convair and the Glenn L. Martin Company. However, their designs were discarded in favor of those produced by an in-house design team under Max Faget. They came up with the idea of a cone-shaped command module, supported by a cylindrical service module containing return propulsion and backup equipment. This would be blasted into space by a rocket designed by Wernher von Braun.

## PROJECT MERCURY

Meanwhile the Mercury program continued. On July 21, 1961, Gus Grissom piloted *Liberty Bell 7* on a second suborbital flight which lasted 15 minutes 37 seconds, reaching an altitude of 118 miles (190 km). On splashdown, explosive bolts blew the hatch off. The capsule filled up with water and sank. Grissom was rescued, though his spacesuit became waterlogged. More suborbital flights had been planned, but were canceled in an effort to catch up with the Soviets who were planning more orbital space shots.

On February 20, 1962, John Glenn made America's first manned orbital flight in *Friendship 7*. The launch had been repeatedly delayed. During the flight, the automatic steering proved faulty. The thrusters failed. The gyroscope indicators were at odds with what Glenn could see for himself and ground control received a signal saying that the lock that held the heat shield in place had opened. If that had been true, the craft would have burned up on reentry.

Scott Carpenter, who orbited the Earth in *Aurora 7* on May 24, 1962, also had problems with the automatic control system and he had to throw the switch to deploy the landing parachute by hand. During his five-hour flight he became the first astronaut to eat solid food in space.

# MAX FAGET

Born in British Honduras, now Belize, Maxime Allen "Max" Faget (1921 - 2004), the son of an American doctor, attended San Francisco, California, Junior College and received a Bachelor of Science degree in Mechanical Engineering from Louisiana State University. He served three years as a naval officer during World War II.

In 1946, he joined the Langley Research Center and worked in the Pilotless Aircraft Research Division. Later he became head of the Performance Aerodynamics Branch. He conceived and proposed the development of a one-man spacecraft, later used in Project Mercury, and was one of the original group of thirty-five, assigned as a nucleus of the Space Task Group to design and carry out the Mercury project.

In February 1962, he became Director of Engineering and Development at the Manned Spacecraft Center for the National Aeronautics and Space Administration in Houston, Texas. He was responsible for technical support of the Gemini and Apollo manned spaceflight programs and advanced studies into space systems, as well as contributing to the later Space Shuttle program.

He was co-author of a textbook, *Engineering Design and Operations of Spacecraft*, and author of a book *Manned Space Flight*. Dr. Faget held joint patents on the "Aerial Capsule Emergency Separation Device" or escape tower, the "Survival Couch," the "Mercury Capsule," and a "Mach Number Indicator."

# MERCURY SPACECRAFT

The Mercury-Redstone Launch Vehicle used for suborbital flights was a single-stage rocket developed from the German V-2. Standing 82 feet (25 m) tall, its engine burned alcohol and liquid oxygen which produced some 80,256 pounds of thrust and speed of 4,225 mph (6,800 kph)—not enough to put the payload into orbit. Originally designed as a ballistic missile by von Braun's team at the Redstone Arsenal in Huntsville, it was modified for the Mercury project by removing the warhead and adding a collar to support the capsule.

Orbital missions used the Atlas LV-3B, a version of the Atlas D, America's first Intercontinental Ballistic Missile. The rocket itself also stood 82 feet (25 m) tall. It was fueled by kerosene and liquid oxygen, developing 356,815 pounds. The first stage had two engines. At liftoff the second stage also fired, with the thrust passing through an opening in the first stage. The second stage continued firing after the first stage separated, putting the capsule into orbit.

The Mercury capsule was 6.8 feet (2 m) long and 6.2 feet (1.9 m) in diameter at the base. It was cone-shaped with a small cylinder on top. The spacecraft weighed between 2,300 and 3,000 pounds (1,043 to 1,360 kg), depending on the mission.

When it sat on top of the booster rocket, there was an emergency escape tower that could lift the capsule off the booster, raising it to a height where the parachutes could open and return the capsule and astronaut to the ground at a safe distance from the launch pad. At the base, there was a heat shield which protected the capsule from the heat of reentry, and retro-rocket thrusters that slowed the capsule down for reentry.

Inside, the astronaut breathed an atmosphere of pure oxygen, with canisters of lithium-hydroxide to remove the carbon dioxide he exhaled. Fans circulated the air and cooled the pilot who wore a pressure suit and helmet for insulation and protection against cabin depressurization.

Walter Schirra orbited the Earth for nine hours on October 3, 1962. The booster engines on the Atlas launch vehicle were misaligned and it rolled clockwise after it left the launch pad. Then his flight suit overheated.

Gordon Cooper made the last and longest Mercury Mission, remaining in orbit for thirty-four hours on May 15 – 16, 1963. He had problems with his suit. There was a power failure on the nineteenth orbit. The carbon dioxide levels in the cabin rose and the temperature soared to 100°F (38°C). The automatic pilot failed and he had to fly the ship by hand during reentry, steering by his knowledge of the stars. He remembered it clearly with a typical aviator's sense of calm:

> *I used my wrist watch for time, my eyeballs out the window for altitude. Then I fired my retro-rockets at the right time and landed right by the carrier.*

## THE KENNEDY ASSASSINATION

Following the Cuban missile crisis, Kennedy again proposed cooperation with the Soviet Union in space. On September 20, 1963, he addressed the United Nations, saying:

> *In a field where the United States and the Soviet Union have a special capacity—in the field of space—there is room for new cooperation, for further joint efforts in the regulation and exploration of space. I include among these possibilities a joint expedition to the Moon. Space offers no problems of sovereignty ... Why, therefore, should man's first flight to the Moon be a matter of national competition?*

Khrushchev was keen, but the US Congress vetoed the idea of a joint mission.

On November 16, 1963, President Kennedy took a tour of the launch site at Cape Canaveral and saw a mock-up of the Saturn rocket that would send Americans to the Moon. On November 21, he praised NASA in a speech in San Antonio, then attended a dinner in honor of Albert Thomas, who had helped bring the Manned Spacecraft Center to Houston. The following day Kennedy was assassinated in Dallas. A week after that the Launch Operations Center at Cape Canaveral was renamed the John F. Kennedy Space Center.

There were fears that the death of President Kennedy would mean the end of the Apollo program. However, it had the opposite effect. It galvanized the whole country. No congressman now dare oppose the legacy of the martyred president, and his successor, President Lyndon B. Johnson, was an even more fervent advocate of the space program.

## PROJECT GEMINI

The two-man Gemini program was already underway. This was to be the bridge between the one-man Mercury and three-man Apollo programs. It was designed to demonstrate that human beings and spacecraft could operate in space for the minimum of eight days needed to undertake a mission to the Moon. The docking procedures needed on a Moon mission had to be practiced. The astronauts would also undertake extravehicular activity (EVA)—spacewalks.

The Gemini spacecraft were carried on a Titan II rocket, a modified ICBM. The crew included astronauts from the Mercury Seven and the 1963 astronaut class, the "New Nine." The Gemini program made some notable achievements. Flying in Gemini 4, Ed White became the first American to make an EVA on June 3, 1965. Again the Soviets had beaten them by a nose with cosmonaut Alexey Leonov undertaking the first spacewalk on March 18, 1965.

---

*I'm coming back in ... and it's the saddest moment of my life.*

Ed White reentering Gemini 4 after his EVA

---

Gemini 5 was the first to use fuel cells to generate power and stayed in space for eight days—the length of time needed to undertake a lunar mission. Gemini 7 made a fourteen-day mission and the first space rendezvous with Gemini 6A. Then Gemini 8 made the first space docking with an unmanned Agena target vehicle.

Buzz Aldrin in Gemini 12 showed that a space traveler could do useful work outside the spacecraft. During a spacewalk that lasted over two hours, he photographed star fields and retrieved a micrometeorite collector.

On June 3, 1965, astronaut Ed White, pilot for the Gemini-Titan 4 spaceflight, performed America's first spacewalk. The extravehicular activity (EVA) lasted 23 minutes as White maneuvered around the spacecraft while attached by a 25-foot umbilical line and a 23-foot tether line. The photograph was taken by Commander James McDivitt from inside the spacecraft early in the EVA over a cloud-covered Pacific Ocean. White enjoyed himself so much he said at the end of the EVA before entering Gemini: "I'm coming back in ... and it's the saddest moment of my life."

# THE GEMINI NINE

On September 17, 1962, a second group of NASA astronauts was announced—known as the "New Nine"—who would pilot the two-man Gemini spacecraft during 1963. Seven of them would be awarded the Congressional Space Medal. Pictured below are, clockwise from top right, Frank Borman, John Young, Thomas Stafford, Pete Conrad, James McDivitt, James Lovell, Elliot See, Ed White, and Neil Armstrong.

• Frank Borman went on to command Apollo 8, which undertook the first manned circumlunar mission. • John Young was the ninth person to walk on the Moon as commander of Apollo 16 in 1972. • Thomas Stafford commanded Apollo 10, which made the first flight of the Lunar Module to orbit the Moon. • Pete Conrad went on to command Apollo 12, becoming the third man to walk on the Moon in November 1969. • James McDivitt commanded Apollo 9 which made the first manned flight of the Lunar Module. • James Lovell was command module pilot on Apollo 8 in December 1968. Then in April 1970, he commanded the ill-fated Apollo 13. • Elliot See was selected as command pilot on Gemini 9 but was killed in a plane crash before the launch. • Ed White was the pilot of Gemini 4 and the first American to walk in space. He was killed in the Apollo 1 fire. • Neil Armstrong went on to command Apollo 11 and became the first man to walk on the Moon in July 1969.

# TITAN II GLV

The Gemini launch vehicle was a modified Titan II missile. It used hypergolic-fuel—a fuel that spontaneously ignites when components are mixed. In the Titan, Aerozine 50 is mixed with the oxidizer nitrogen tetroxide, giving 469,850 pounds of thrust. The rocket stood 108 feet (33 m) tall and was 10 feet (3 m) in diameter. Modifications to Titan II for use as the Gemini Launch Vehicle included:

- A malfunction detection system was added to detect and transmit booster performance information to the crew.
- Backup flight control system was added to provide a secondary system if the primary system fails.
- Radio guidance substituted for inertial guidance.
- The second stage propellant tanks were lengthened for longer burn time and unnecessary vernier engines and retro-rockets were removed.
- New second stage equipment truss added with a new second-stage forward oxidizer skirt assembly.
- Modifications were made to the tracking, electrical, and hydraulics systems to improve reliability.
- The propellants were chilled to slightly improve vehicle performance, also allowing for more fuel to be carried.
- The first stage engine thrust was reduced slightly to cut down on vibration and the g-force on the astronauts.

A time-lapse image shows the launch tower falling away as a Titan II GLV rocket launches Gemini 10 into space, July 18, 1966. This was the eighth manned Gemini flight and lasted three days.

# THE GEMINI CAPSULE

The capsule used on the Gemini missions was essentially an upgraded Mercury capsule and built by the same contractor, McDonnell Aircraft. At 8 feet 5 inches (5.61 m) long and 10 feet (3 m) wide, it had been enlarged to carry two men. It was modular in design, so as to make it easier to repair. It also used solid-state electronics. The launch weight varied from 7,100 to 8,350 pounds (3,220 to 3,790 kg).

The astronauts sat side by side in the capsule. The commander, known as the command pilot, sat in the left seat, while the pilot sat in the right seat. The crew sat in ejection seats to provide the astronauts with a means of escape, at least at low altitudes.

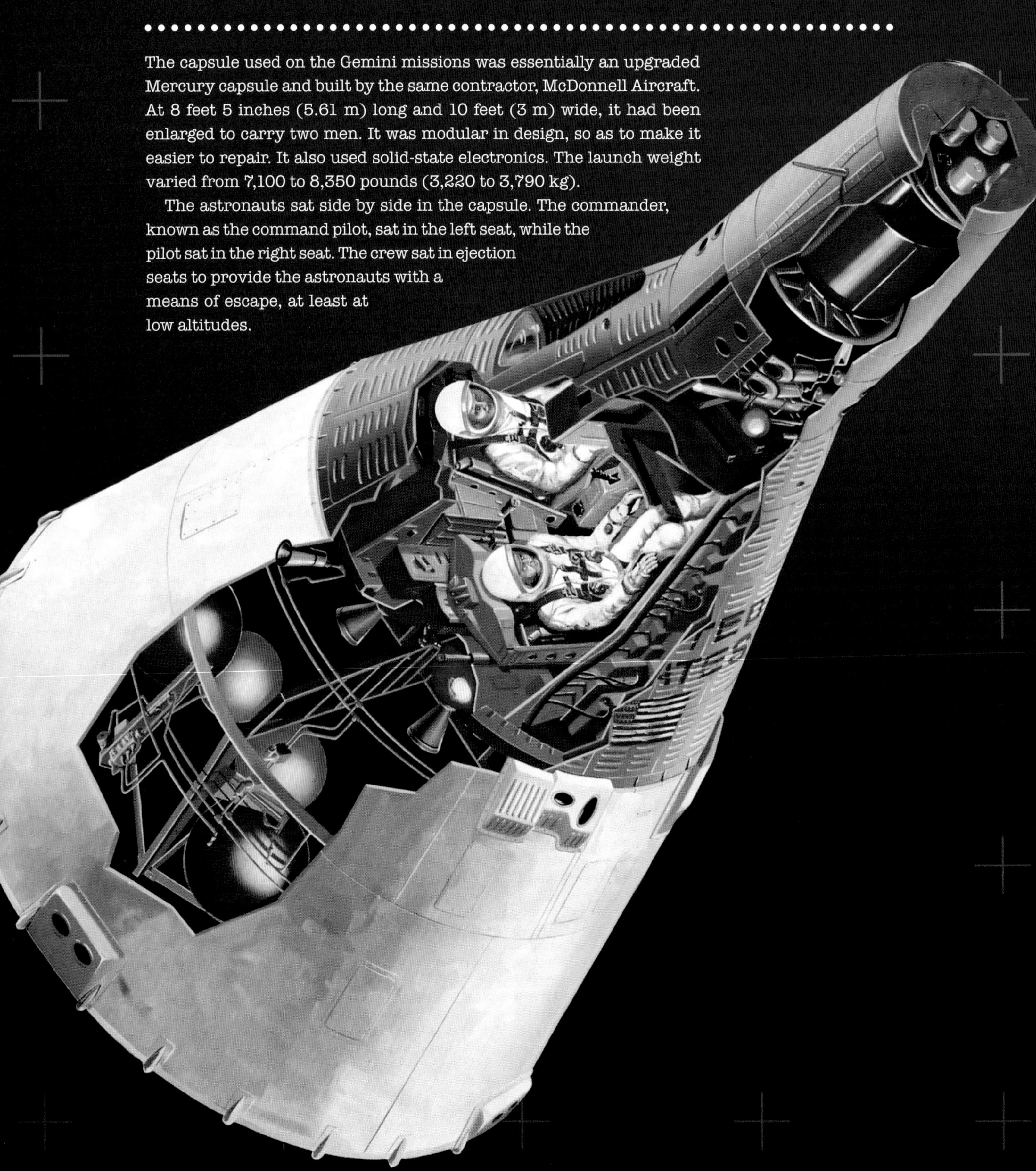

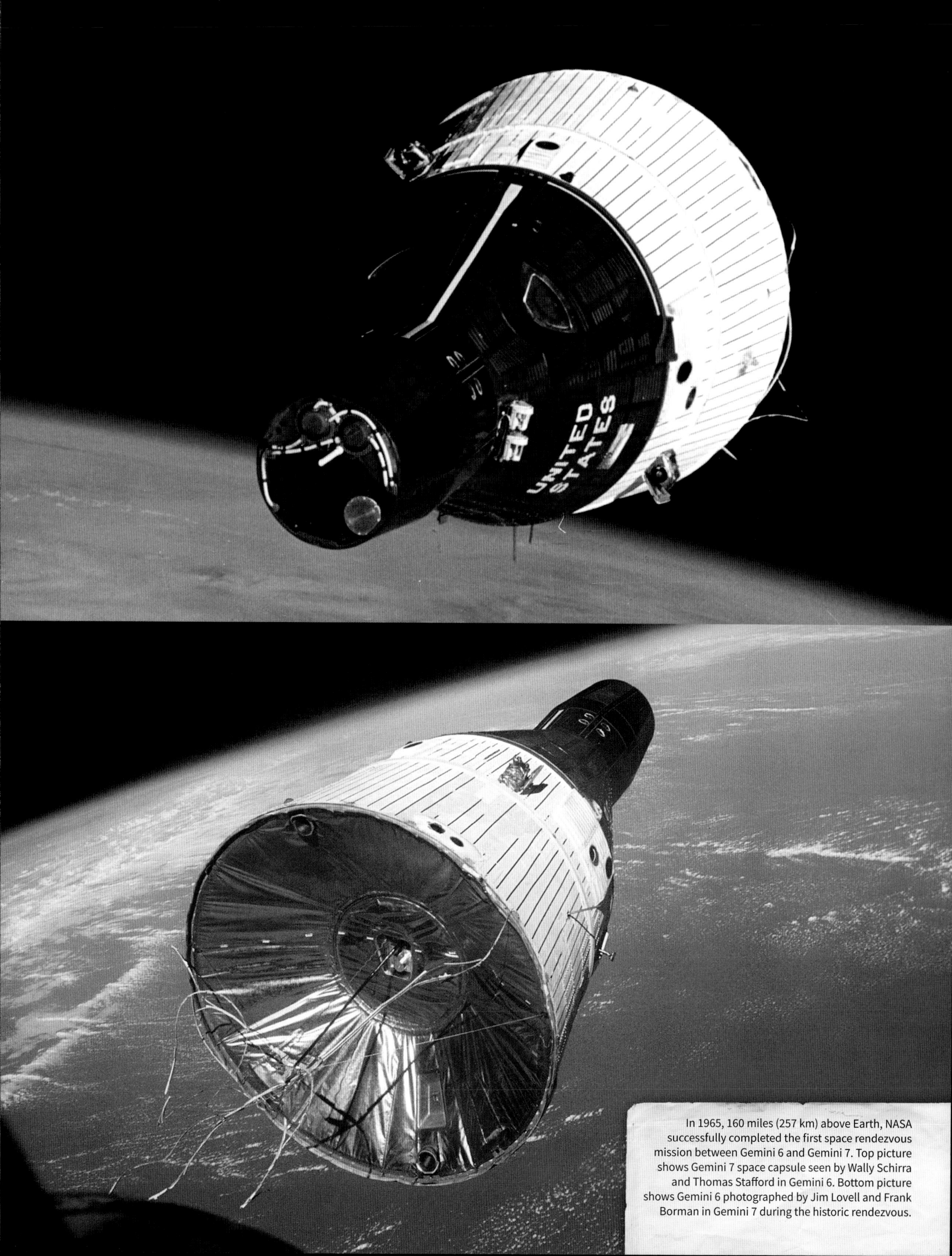


In 1965, 160 miles (257 km) above Earth, NASA successfully completed the first space rendezvous mission between Gemini 6 and Gemini 7. Top picture shows Gemini 7 space capsule seen by Wally Schirra and Thomas Stafford in Gemini 6. Bottom picture shows Gemini 6 photographed by Jim Lovell and Frank Borman in Gemini 7 during the historic rendezvous.

CHAPTER 7

# GOD SPEED TO APOLLO 1

## APOLLO 1

On September 1, 1963, George Mueller replaced Brainerd Holmes as head of the Office of Manned Space Flight. He immediately commissioned a report into whether Apollo could reach the Moon by 1970 and beat the Russians. It reported that the chances of success were just ten percent. Mueller and Robert Seamans decided to ditch the report in case it damaged NASA in the eyes of Congress.

To speed up the Apollo program, Mueller also ditched the meticulous item-by-item testing that von Braun and his German colleagues had brought from the V-2 development team and introduced "all-up testing," where the whole rocket would be assembled and tested in one go.

"It sounded reckless, but George Mueller's reasoning was impeccable," said Wernher von Braun. "In retrospect it is clear that without all-up testing the first manned lunar landing could not have taken place as early as 1969."

### DEVELOPING THE SATURN MOON ROCKET

Wernher von Braun had long held the ambition to put a man on the Moon. He knew the secret to this was multi-staged liquid-fueled rockets. The staging was vital. As each stage was exhausted, it fell away, making the rocket lighter.

To lift the heavy payloads required, he turned to the concept of clustering. By strapping together the booster rockets from the Jupiter missile, he intended to build a "Super-Jupiter" called the Juno V.

But von Braun was on the Army payroll, and they wanted missiles not Moon rockets. Nevertheless, he persisted. He made a first stage with eight Rocketdyne engines under eight propellant tanks from his Redstone rocket. These were clustered around a Jupiter propellant tank. The second stage used pieces from Titan and Centaur series rockets. Together they could put a payload into low-Earth orbit and it became known as

# GEORGE MUELLER

Unable to afford a course in aeronautical engineering, George Edwin Mueller (1918 – 2015) studied electrical engineering at the Missouri School of Mines and Metallurgy, now Missouri University of Science and Technology. He got a television fellowship at Purdue University and built a TV transmitter for the campus there. At Bell Labs during World War II, he worked on airborne radar and semiconductors.

After obtaining a PhD in physics at Ohio State University, he worked on radar systems for the Titan rocket and became increasingly involved in missile systems. He then took a pay cut to join NASA in 1963. His job was to put the Apollo program back on schedule and reorganized the Manned Space Flight Center at Huntsville despite the objections of Wernher von Braun.

He formed the Apollo Executive Group with the presidents of the main contractors. To save time and money he developed "all-up testing," which tested the whole assembly together rather than search for problems with individual components.

After the Apollo 1 fire, James Webb at NASA grew wary of Mueller, but said: "Even if I wanted to, I couldn't fire him because he was manager of our successful Apollo project, and one of the ablest men in the world ... The last thing I wanted was to lose him, but I also had another desire, which was not to let his way of working create too many difficulties."

After the success of Apollo, Mueller championed Skylab and the Space Shuttle. With the termination of the Apollo program he quit NASA and went into private industry.

Saturn as it was the one after Jupiter in the order of planets going outward from the Sun.

However, this was of little use to the Army. The Juno V, then known as the Saturn 1-A, was canceled and von Braun's team at the Army Ballistic Missile Agency at Huntsville was transferred to NASA. They began redesigning the Saturn 1-A, using the new F-1 engines being developed by the Air Force that would give over 1.5 million pounds of thrust.

NASA had been considering building a new rocket from scratch named Nova to take men to the Moon, but it was clearly a better option to use the tried-and-tested technology of the Saturn.

Aerospace engineer Abe Silverstein chaired the Saturn Evaluation Committee, better known as the Silverstein Committee. He came up with two fast-track solutions. These were Saturn A and the Saturn B, which used the technology from the Jupiter, Titan, and Centaur missiles. Then there would be the Saturn C series that would use the new F-1 and the J-2 engines.

Silverstein considered that only the three-stage Saturn C-5 would be capable of getting a manned mission to the Moon and it was renamed the Saturn V. Although it only existed on paper, in January 1962, NASA announced that the Saturn V would be used on the Apollo mission. While it was in development, NASA would build the smaller Saturn C-1 as a test vehicle since its lower stages used tanks from the Redstone and Jupiter missiles. The two-stage Saturn C-1 was renamed Saturn 1.

## SATURN 1

As on the Juno V, the first stage of the Saturn 1 used H-1 engines with eight Redstone tanks clustered around a Jupiter tank. Four of the Redstone tanks carried RP-1, four carried liquid oxygen, with more liquid oxygen in the Jupiter tank. The H-1 engines would be configured with a square of four around a central cluster of four. The outer four gimbaled for steering and the entire arrangement gave a thrust of 1.5 million pounds, more than ten times that of the Jupiter-C rocket that had launched Explorer 1 three years earlier.

### The F-1 Rocket Engine

Built by the Rocketdyne Division of Rockwell International, the F-1 used RP-1 (a jet-fuel type of kerosene) and liquid oxygen as the propellants. Its 2,500-pound turbopump pumped in the propellants at 42,500 gallons a minute. Its first static firing was in March 1959 and it was delivered to NASA in October 1963. The F-1 engine designed to carry men to the Moon would have to be in an order of magnitude more powerful than the existing engines used in ballistic missiles. Combustion instability caused the engine to blow up. Twenty of the first forty-four tests failed. It took seven years before it passed its flight-rating test. The result was the most powerful single-nozzle liquid-fueled rocket engine ever flown. The Saturn V had five of them.

F-1 ENGINE

18.5 FT

12.2 FT

The second stage, S-IV, was fueled by liquid hydrogen and liquid oxygen. To save weight these had to be stored in adjoining tanks, even though they were stored at vastly different temperatures, presenting problems for the constructors. Onboard computers would be used to control the ascent, a concept that proved crucial on the larger Saturn V.

On October 27, 1961, the first Saturn 1 was launched successfully, carrying a dummy second stage. Only five months since President Kennedy had made his historic announcement, the first prototype Moon rocket had soared skyward—and this was before John Glenn had even orbited the Earth. Nine more Saturn 1 flights were made.

It was then superseded by the Saturn 1B, which carried an improved second stage called the S-IVB. This would be used as the third stage of the Saturn V as it could crucially reignite in space. On February 26, 1966, a Saturn 1B carried prototype Apollo Command and Service Modules aloft. Two more unmanned suborbital test flights were made that year.

## DISASTER STRIKES AS-204

Following the successful Mercury and Gemini programs, there was good reason for optimism about Apollo's manned missions to the Moon. Three unmanned test flights (AS-201, AS-202, and AS-203) carrying the Command and Service Modules had been successful. But disaster struck during preparation for the launch of AS-204, the first manned orbital flight, and the mission never flew.

On January 27, 1967, the three-man crew—Virgil "Gus" Grissom, a 40-year-old veteran of the Mercury program, 36-year-old Ed White who made the first US spacewalk with Gemini, and 31-year-old newcomer Roger B. Chaffee—were in their capsule on top of a twenty-story Saturn 1B rocket on Launch Complex 34 of the Kennedy Space Center four weeks before their launch date. They were breathing pure oxygen as they would during the mission.

There were problems with the radio communication, but at 6:31 p.m. Deke Slayton, who was monitoring the

J-2 ENGINE

11.1 FT

6.8 FT

## The J-2 Engine

The J-2 engine was developed by Rocketdyne in 1960 and was fueled by liquid hydrogen and liquid oxygen, producing 1.5 million pounds of thrust. It was so versatile that it was used for both the second and third stages of the Saturn V. It was the first manned booster engine that used liquid hydrogen as a fuel and the first large booster engine designed to be restarted multiple times during a mission. To do this, the liquid propellants had to be at the "bottom" of the tank, with the gas above them, otherwise gas would be fed into the engine instead of liquid and the engine would fail to reignite. Two small solid rockets were housed in external pods. Immediately before the J-2 engine had to restart, they were ignited, throwing the rocket forward and forcing the fuel back into the propellant pumps. The supply of liquid hydrogen to the upper-stage J-2 engines was also a problem. Liquid hydrogen boils at −423°F (−252°C). This can cause bubbles and vapor pockets, and the tiny molecules of hydrogen can leak through piping suitable for heavier fuel.

test, thought he heard someone say "fire." He looked at the CCTV screen and saw that the capsule window was glowing. Another camera picked up Ed White's hand reaching for the hatch bolts. Then Slayton heard Chaffee say: "We've got a fire in the cockpit."

Skip Chauvin, who was conducting the test, called for the electricity to be shut down. Then Chaffee's voice was heard again, saying: "We've got a bad fire … Let's get out … We're burning up."

After just twelve seconds, the capsule exploded. A sheet of flame burst out of the side and the temperature soared to 2,500°F (1,371°C). There was nothing the technicians could do for the crew as the hatch door was sealed from the inside. Instead they sprayed coolant on the nozzles under the rocket engines, otherwise the whole thing would have gone up, killing everyone on the pad.

The heat and smoke were intense. Finally, the hatch was released. Inside the three astronauts were dead. Witnesses could not believe what they saw. They thought that the crew had been burned alive. In fact, they had suffocated from the fire's toxic gases. All operations were halted while those on hand made a note of everything they had seen and heard.

## A ROUGH ROAD LEADS TO THE STARS

James Webb appointed a review board to oversee the investigation. Although they knew that using pure oxygen in the cockpit was risky, it had been used on Mercury and Gemini spacecraft. However, in Apollo there were large amounts of nylon and Velcro which were flammable. There seemed to have been an electrical short in the lower equipment bay near Gus Grissom's left foot. This created a spark. In an atmosphere that was 100 percent oxygen at a pressure of around 1.5 times that of the atmosphere, fire spread rapidly, creating an explosion.

The fire damaged NASA in the eyes of the American people, but the astronauts themselves, who had all been fighter pilots and test pilots, accepted that risk was part of their business. The Soviet space program had accidents too—one remarkably similar to the AS-204 disaster.

On March 23, 1961, 24-year-old Valentin Bondarenko was training in a capsule pressurized with pure oxygen. He had removed the biosensors from his torso, using cotton wool soaked in alcohol. When he tossed it aside, it hit a hot plate and ignited. The fire flashed

The burnt-out interior of Apollo 1 after the fire that killed astronauts Gus Grissom, Ed White, and Roger Chaffee.

over onto his suit. By the time the hatch was opened, the skin over his entire body had been charred with the exception of the soles of his feet which were protected by his boots. He died eight hours later. But the Soviets managed to conceal his death until 1980.

A Congressional investigation discovered that North American Aviation had been awarded the contract to build the Apollo capsule over the Martin Company. James Webb explained that the astronauts themselves opted for North American as they had made the X-15 rocket-powered airplane for NASA, which set the world speed recorded for a manned aircraft. The X-15 pilots had included Neil Armstrong. Grissom himself had favored North American.

Ed White was buried at West Point; Gus Grissom and Roger Chaffee were interred at Arlington National Cemetery. They left wives and children. They sued North American and won over $600,000 in damages.

*There's always a possibility that you can have a catastrophic failure, of course; this can happen on any flight; it can happen on the last one as well as the first one. So you just plan as best you can to take care of these eventualities, and you get a well-trained crew and you go fly.*

Gus Grissom, December 1966

## WHAT HAPPENED TO APOLLO 2 AND 3?

In honor of Astronauts Virgil I. Grissom, Edward H. White II, and Roger B. Chaffee, on April 24, 1967, NASA's Office of Manned Space Flight announced that AS-204 would be renamed as Apollo 1.

The earlier, unmanned Apollo Saturn IB missions AS-201, AS-202, and AS-203 were not given "Apollo" flight numbers and no missions were named "Apollo 2" and "Apollo 3." The next mission flown, the first Saturn V flight (AS-501, Apollo Saturn V No. 1), skipped numbers 2 and 3 to become Apollo 4.

"Apollo 1" although it was never launched, was officially recorded by NASA as "First manned Apollo Saturn flight—failed on ground test."

The unmanned AS-201 launched from the Kennedy Space Center, 1966.

## ED WHITE

After graduating from West Point, Edward Higgins White II (1930 - 67) was commissioned in the USAF. He studied Aeronautical Engineering at the University of Michigan and trained as a fighter pilot. Logging more than 3,000 flying hours, he attained the rank of lieutenant colonel.

As pilot of Gemini 4, he was chosen to make the first spacewalk on June 3, 1965. He enjoyed it so much, that he had to be ordered back into the spacecraft. He lost a spare glove during the EVA and his co-pilot James McDivitt had trouble resealing the hatch after White reentered the spacecraft.

His second flight was to have been onboard Apollo 1, scheduled for launch on February 21, 1967, but he was killed in the test of the spacecraft. He left behind his wife, Patricia, and two children.

## GUS GRISSOM

One of the original Mercury Seven, Lieutenant Colonel Virgil I. "Gus" Grissom (1926 - 67) was the second American in space. Still in high school when America joined World War II, he volunteered as an aviation cadet. He married Betty Moore as the war drew to its end. After being discharged from the Army, he studied Mechanical Engineering at Purdue University.

He re-enlisted during the Korean War, winning the Distinguished Flying Cross and the Air Medal with an oak leaf cluster. In 1955, he studied Aeromechanics at the US Air Force Institute of Technology in Dayton, Ohio, before training to be a test pilot. This earned him an invitation to join the space program. His first space-flight lasted 15 minutes 37 seconds. On splashdown the capsule filled with water and was lost. He went into space again in Gemini 3.

Before his death, he was a candidate to be the first man on the Moon. He was survived by his wife and two sons.

## ROGER CHAFFEE

A Naval Reserve Officers Training Corps took Roger Bruce Chaffee (1935 - 67) to the Illinois Institute of Technology and Purdue University where he also obtained a private pilot's license. After graduation, he was commissioned as an ensign in the US Navy and trained as a pilot. He then served as an officer in Heavy Photographic Squadron VAP 62 which took pictures of the launch sites on Cuba during the Cuban missile crisis and attained the rank of lieutenant commander, logging more than 2,300 flying hours. He also studied Reliability Engineering at the US Air Force Technology.

In 1963, he was selected to be among NASA's Astronaut Group 3 alongside Buzz Aldrin, and served as capsule communicator on Gemini 3 and 4. Three years later he was selected as a crew member on Apollo 1 and flew out to the North American Aviation plant in California to see the spacecraft. Chaffee's widow received $100,000 as part of a publishing deal.

...w of Apollo 1 who tragically died in ...on the launch pad, left to right: Ed ...us Grissom, Roger Chaffee.

## God Speed to the Crew of Apollo 1

When Launch Complex 34 was decommissioned, the launch platform was left standing as a memorial to the crew of Apollo 1. It carries two plaques. One reads:

LAUNCH COMPLEX 34

Friday, 27 January 1967

1831 Hours

Dedicated to the living memory of
the crew of the Apollo 1

U.S.A.F. Lt. Colonel Virgil I. Grissom

U.S.A.F. Lt. Colonel Edward H. White, II

U.S.N. Lt. Commander Roger B. Chaffee

They gave their lives in service to their country in the ongoing exploration of humankind's final frontier. Remember them not for how they died but for those ideals for which they lived.

The other plaque in remembrance of the crew reads:

IN MEMORY
OF

THOSE WHO MADE THE ULTIMATE SACRIFICE
SO OTHERS COULD REACH THE STARS

AD ASTRA PER ASPERA
(A ROUGH ROAD LEADS TO THE STARS)

GOD SPEED TO THE CREW
OF

APOLLO 1

*If we die, we want people to accept it. We hope that if anything happens to us it will not delay the program. This conquest of space is worth the risk of life.*

Gus Grissom

1017
1019
1013
1012
1015
1011
1016
1021
STAY CLEAR

PART THREE

# HOW TO BUILD A SPACECRAFT

*THE DREAM OF YESTERDAY IS THE HOPE OF TODAY AND THE REALITY OF TOMORROW.*

ROBERT H. GODDARD
AMERICAN ROCKET PIONEER

# CHAPTER 8

# SATURN V

To land a man on the Moon, NASA intended to use a technique called Lunar Orbit Rendezvous. A main spacecraft and a smaller lander would be put into lunar orbit. The lander would then detach and descend to the surface of the Moon. When the mission was complete, the lander would lift off and rendezvous with the main spacecraft. The lander could then be discarded while the main spacecraft returned to Earth. This technique gave considerable savings in weight and was first proposed in 1919 by the Soviet engineer Yuri Kondratyuk.

## TALLER THAN THE STATUE OF LIBERTY

Despite the advantages of Lunar Orbit Rendezvous, the rocket needed to send men to the Moon would have to be massive. It would stand 363 feet (110.6 m) tall—the height of a thirty-six-story building and 60 feet (18 m) taller than the Statue of Liberty. Fully fueled, it weighed 3,100 tons, as much as 500 elephants. It developed 7.6 million pounds of thrust at launch, creating more power than eighty-five Hoover Dams. It could lift 130 tons into Earth orbit. That's the weight of ten school buses. And it could launch 50 tons, or four school buses to the Moon. Although each rocket would take a year to build, its working lifetime would be a matter of minutes, guzzling enough fuel to take the average automobile around the world 800 times. Each would cost $100 million, equivalent to $650 million in 2018. Once the launch was over, the pieces fell into the ocean.

The infrastructure needed to handle it was huge. The Vehicle Assembly Building (VAB) where the stages were stacked was the size of a cathedral and the crawlers used to carry the rocket out to the launch pad were two of the largest tracked vehicles ever made.

> *Our two greatest problems are gravity and paperwork. We can lick gravity, but sometimes the paperwork is overwhelming.*
>
> Wernher von Braun

## FIRST STAGE S-IC

Built by Boeing, the first stage of the Saturn V was five F-1 engines clustered together—giving the rocket the designation V, or a Roman five. They were gimbaled to steer the rocket. In the first 2 minutes 30 seconds they burned enough kerosene and liquid oxygen to fill an Olympic-size swimming pool, taking it from zero to 5,300 mph (8,530 kph) and a height of 35 miles (56 km).

The first stage was 138 feet (42 m) tall. Above the F-1 engines stood two tanks, 33 feet (10 m) wide, stacked one on top of the other. The lower tank

# YURI KONDRATYUK

Born in Poltava in the Ukraine, then part of the Russian Empire, Yuri Vasilievich Kondratyuk's real name was Aleksandr Ignatyevich Shargei (1897 – 1942). His father had studied physics and mathematics at Kiev University and, from an early age, Kondratyuk was fascinated by his books. He excelled at school and enrolled in the Great Polytechnic in St. Petersburg. At the outbreak of World War I he was drafted into the Russian Army as a warrant officer. Serving on the Caucasus Front, he filled four notebooks with his ideas for interplanetary flight. His notes included an outline of the Lunar Orbit Rendezvous technique used in the Moon landing.

Following the Russian Revolution, he left the army, but as a former officer in the Tsarist army he was in constant danger of being arrested by the Bolshevik authorities as an "enemy of the people." In hiding, he managed to obtain forged identity papers in the name of Yuri Vasilievich Kondratyuk, who had died.

Making his living as a mechanic, he wrote *The Conquest of Interplanetary Space* which also outlined the gravitational slingshot effect used by NASA during the Apollo missions. Tackling more practical problems, he designed a huge wooden grain elevator, constructed without a single nail as metal was in short supply. The Soviet secret police concluded that he had left out the nails because he intended it to collapse. Charged with being a saboteur, he was sentenced to three years in the Gulag. However, because of his evident talents, this was commuted to exile.

He designed a wind power generator in the Crimea. Sergei Korolev, head of the Soviet rocket research group, offered him a job, but he turned it down, fearing security checks would reveal his true identity. When Korolev was charged with treason for wasting time designing spacecraft, Kondratyuk decided to get rid of his notes. He left them with a trusted neighbor who took them to the US when she escaped from the Soviet Union after World War II.

During the war, Kondratyuk joined the Red Army. He was listed as missing in action in the winter of 1941 – 42 and almost certainly died.

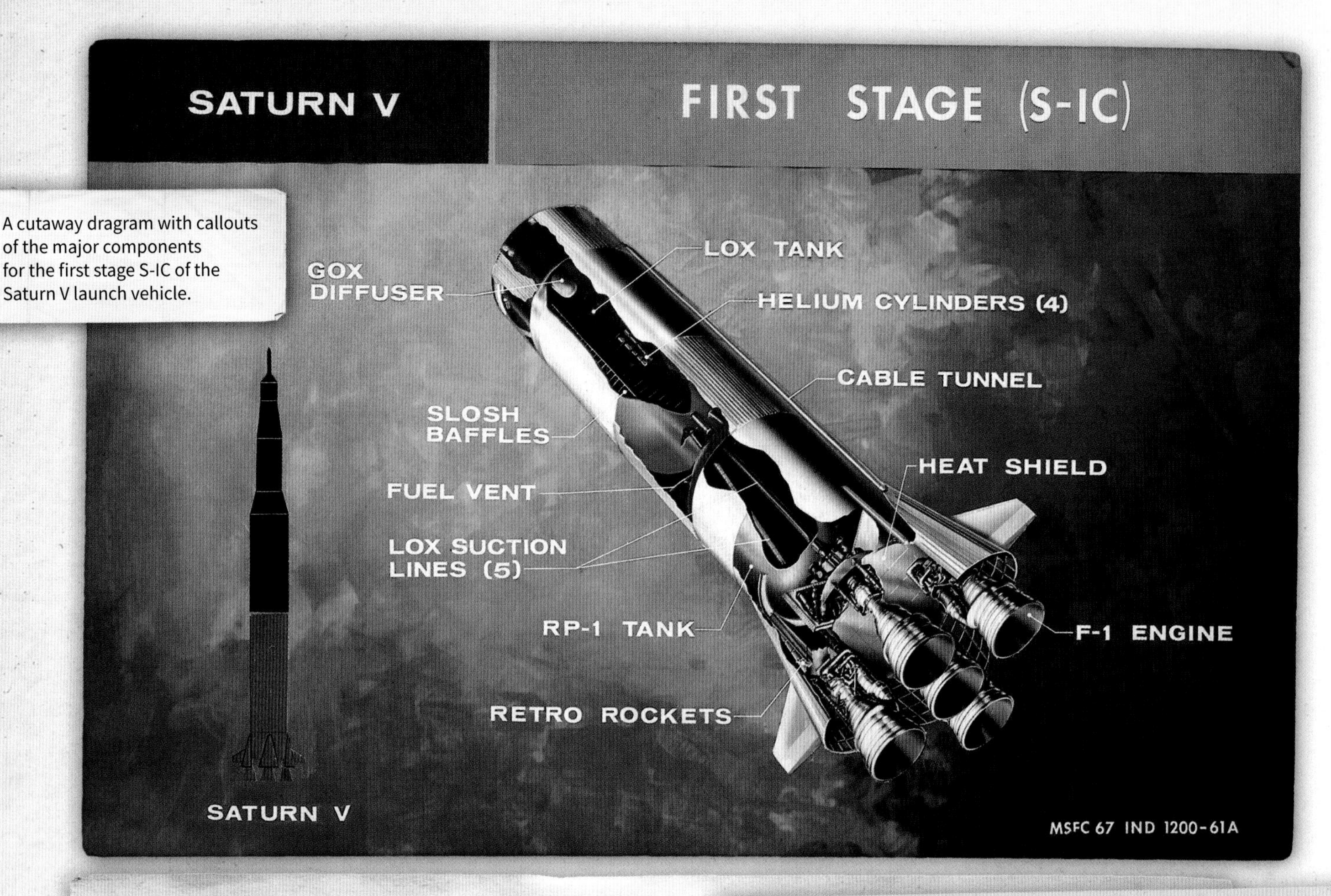


A cutaway dragram with callouts of the major components for the first stage S-IC of the Saturn V launch vehicle.

## Vehicle Assembly Building (VAB)

The Saturn V was built by numerous contractors and subcontractors at factories stretching from California to Alabama. Components were brought by custom-made craft and barge to Cape Kennedy where they were assembled in the VAB, which is the largest single-story building in the world.

During construction, 4,225 pilings were driven down 164 feet (50 m) to bedrock. The foundation required 30,000 cubic yards of concrete and 98,590 tons of steel were used to make the building. It covers eight acres, is 525 feet (160.3 m) tall, 716 feet (218.2 m) long, and 518 feet (157.9 m) wide, enclosing 129,428,000 cubic feet of space.

To assemble the rocket, there were seventy-one cranes and hoists, including two 250-ton bridge cranes. On the east and west sides, four high-bay doors, each designed to open up to 456 feet (139 m) in height, allowed rollout of the rockets mounted atop launch umbilical towers. The VAB was later modified to accommodate the Space Shuttle.

NASA's Vehicle Assembly Building at Kennedy Space Center in Florida.

held 211,338 gallons (800,000 liters) of RP-1 jet-aircraft grade kerosene. The upper tank contained 343,423 gallons (1.3 million liters) of liquid oxygen at –297°F (–183°C). Almost anything will burn in liquid oxygen, or LOX, so it had to be handled with particular care. Not even a fingerprint could be left inside the tank holding it. The LOX was carried down via five large, insulated ducts through the RP-1 tank to the engine.

The F-1 engine was the most powerful rocket engine ever built. Each burned fifteen tons of kerosene a second, producing 180 million horsepower that could lift 680 tons. They were put through their paces on test stands in the Mojave Desert. Beneath the engine bells were giant flame buckets to direct the thrust outward to stop it blasting away the foundations. The vibrations could be felt a mile away. In adverse weather conditions, the shock wave stayed at ground level damaging houses miles away.

The first stage of the Saturn V was simply meant to be a scaling up of technology already used in earlier rockets. However, this did not always work. The speed the propellant was injected into the combustion chamber led to dangerous instabilities which, on June 28, 1962, led to a test rig blowing up.

In those days, there was no computer modeling, so it was necessary to rerun the firing sequence repeatedly until the problem was solved. After NASA had lost two engines, a new method of checking was employed. They detonated small bombs inside the engine to set off the instability. Then they could see whether the modifications made had dampened the dangerous oscillations. If not, they could close the engine down before any catastrophic damage was done.

The problem was with the way the propellant was injected into the chamber. Copper baffles on the face of the injector dampened the oscillations within 400 milliseconds, making the engines safe enough for manned flight.

## The Crawler-Transporters

The crawler-transporters at the Kennedy Space Center are 131 feet (40 m) long and 114 feet (35 m) wide. Built by the Marion Shovel Company in Marion, Ohio, they were shipped to Kennedy Space Center in 1964 where final assembly took place. The crawlers started their service in 1965 to support the Apollo program. The first practical use of a crawler was in August 1967, when the first Saturn V rocket for Apollo 4, an uncrewed mission, was transported to Launch Pad 39A. Before that, rocket stages were erected and assembled at the pad. Altogether the crawlers carried nineteen Saturn rockets to their pads until 1975, supporting the Skylab and Apollo-Soyuz Test Project missions as well as Apollo. Between 1981 and 2011, the crawlers transported Space Shuttles 135 times on their 4.2-mile trek from the VAB to Launch Pad 39A or 39B and back, at a maximum speed of one mile an hour loaded, and about two miles an hour unloaded, burning 150 gallons of diesel oil per mile during their super slow ride. They were still in use after fifty years and in 2016 were upgraded.

The crawler-tranporter taking Apollo 14 to the launch pad at Kennedy Space Center, 1971.

## SECOND STAGE S-II

The first stage cut out at 220,000 feet (67,000 m) and dropped away. The rocket and spacecraft now weighed just 1,000 tons, less than a third of its launch mass. The second stage then carried it up to 610,000 feet (186,000 m) at four miles a second (14,400 mph or 23,175 kph).

It also had five engines, but these were J-2s using liquid hydrogen and liquid oxygen as fuel. Like in the first stage, the engine were arranged with four engines in a square around a central fixed one. The outer engines were gimbaled for steering. The J-2 engines were more efficient, but did not have the raw power of the F-1. Together they could lift 520 tons.

The second stage was the largest cryogenic rocket ever built. It had to be shipped from Los Angeles through the Panama Canal for testing. The issue for the engineers was that the second stage was the last stage to be designed, so that any weight gain in the rest of the launch vehicle and spacecraft had to be lost in the second stage.

One way to save weight was to abut the liquid oxygen and liquid hydrogen tanks with a single bulkhead between them. The problem was that the temperature of the liquid oxygen was at –297°F (–183°C), while the liquid hydrogen was –423°F (–253°C). Thickening the aluminum wall between the two tanks added weight. Instead it was decided to use a honeycomb insulation.

However, when the liquid propellants were pumped in, air pockets trapped in the bonding glue would freeze, lifting off the insulation. Fortunately, the factory was at Seal Beach in California and some of the engineers were keen surfers. Surfboards were made from a honeycomb of polyurethane or polystyrene foam covered with layers of fiberglass cloth, and polyester or epoxy resin. They came up with a way to cut grooves in the insulation, then purge any remaining air with helium that only freezes at –458°F (–272°C). This saved four tons.

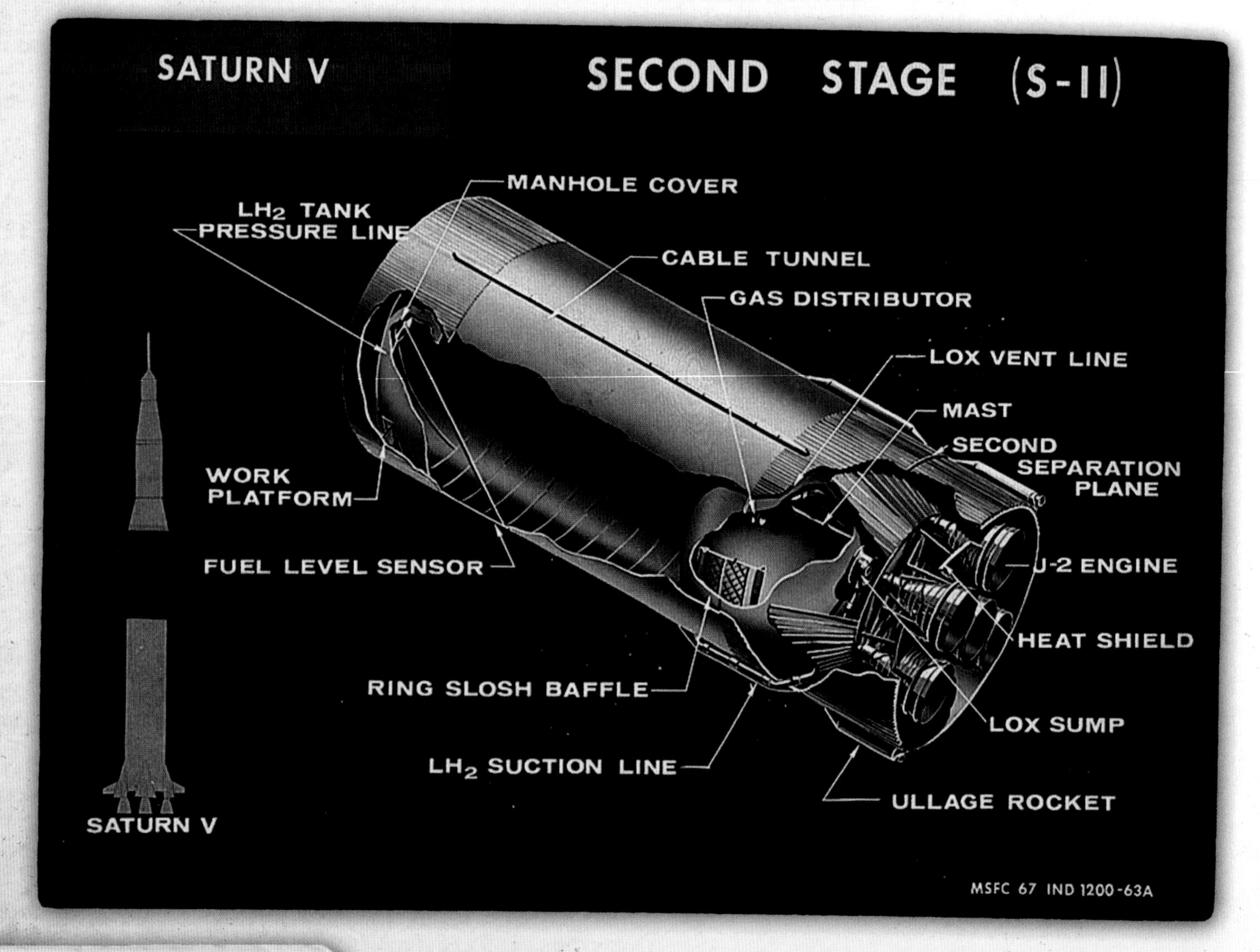


A cutaway diagram with callouts of the major components of Saturn V second stage S-II.

## THIRD STAGE S-IVB

The third stage was the S-IVB second stage previously used in the Saturn 1B. It would carry the spacecraft into Earth orbit, then reignite to propel it on toward the Moon. Four solid-stage ullage rockets were arranged around it to push the propellant into the pumps.

At the top of the third stage was a 3 feet (91 cm) high ring, housing what was called the Instrumentation Unit. This was a gyroscopically stabilized platform connected to a computer, giving the rocket its own autonomous guidance system. There were three processors working in parallel. If one of them deviated from the flight plan, the other two would take over. Many features were over-engineered in this way. As the Saturn V had 5.6 million parts, even if their reliability was 99.99 percent, in any launch you could expect 560 failures, so duplicate systems were employed to compensate. In emergencies, this could be overridden by the mission commander in the Command Module.

Wernher von Braun called the Instrumentation Unit "the Saturn's most critical stage." It meant that the Saturn V could be given an "all-up test," testing the whole assembly together, rather than each element separately. It also gave NASA the confidence to send men to the Moon on only the third Saturn V ever flown. Von Braun had originally planned at least ten unmanned test flights. George Mueller overrode this, insisting that all the stages be tested together in one stack. It was a high-risk strategy but NASA was determined to beat the Soviets.

A cutaway drawing showing the third stage S-IVB of the Saturn V launch vehicle.

# CHAPTER 9

# THE CSM

The Command Module (CM) had to carry everything that three men needed to live for three weeks—power, temperature control, shielding from cosmic rays, food, water, air, hygiene facilities, and waste disposal. It also needed all the systems necessary to take it on the half-million-mile flight from the Earth to the Moon and back—propulsion, thrusters for steering and attitude control, deep-space communications, navigation and guidance, a heat shield for reentry, and parachutes to slow the craft for splashdown.

All this would make the craft so heavy that it would be impossible to slow it down sufficiently on reentry and it would burn up. However, a solution had already been pioneered on Gemini. Most of these systems could be carried in a separate Service Module (SM) slung under the capsule, which could be jettisoned shortly before reentry. The lightweight capsule, or CM, could then make the reentry on its own.

During launch and throughout the bulk of the mission the spacecraft would be known as the Command and Service Module, or CSM.

Like its Mercury and Gemini forerunners, the CM would be conical. It was 10 feet 7 inches (3.2 m) high and 12 feet 10 inches (3.9 m) in diameter. It weighed 13,000 pounds (5,896 kg), which was reduced to 11,700 pounds (5,307 kg) on splashdown, as the propellant for the reaction control system—monomethyl-hydrazine oxidized by tetroxide—weighed some 270 pounds (122 kg).

The volume of the crew compartment was 210 cubic feet (5.9 cubic meters), about the same as the interior of a large family automobile. However, in zero gravity, this was not as cramped as the Mercury or Gemini capsules.

## THE WINDOWS IN THE CSM

There were five windows. The main hatch door had a porthole 9 inches (23 cm) in diameter over the central couch, though on some missions this was replaced by a scientific air lock. There were two large windows 13 inches (33 cm) square, one either side of the hatch. Two triangular rendezvous windows about 8 by 13 inches (20 by 33 cm) pointed forward which were used during docking.

The windows were double-skinned. The inner windows were made of tempered silica glass with quarter-inch (6.3 mm) thick double panes, separated by a tenth of an inch (2.5 mm). The outer windows were made of amorphous-fused silicon with a single pane seven-tenths of an inch thick (18 mm).

Each pane had an anti-reflecting coating on the external surface and a blue-red reflective coating on the inner surface to filter out most infrared and all ultraviolet rays. Some missions used windows made from quartz to allow UV photography. The outer window glass had a softening temperature of 2,800°F (1,538°C) and a melting point of 3,110°F (1,710°C). The inner window glass had a softening temperature of 2,000°F (1,093°C). They had shades made of aluminum sheet with a non-reflective inner surface.

During launch, reentry, and other major maneuvers, the crew would lie on three couches. They would be facing upward in the direction of travel during launch. At their feet would be the navigation and guidance system, including the Apollo guidance computer, a sextant for optical sightings, and the communications equipment. Either side were lockers to carry Moon samples on the return journey.

## THE DISPLAY CONSOLE

Immediately in front of the couches and within easy reach was the main display console which was built around the docking tunnel. This accommodated the main flight controls and instruments—more than 400 switches, read-outs, warning lights, and alarms. There were more switches and circuit-breakers down the sides of the couches.

The rest of the interior walls were covered in storage bays for clothes, food, medical and hygiene kits, waste-containers, survival gear, cameras, and the containers of lithium hydroxide needed to take the carbon dioxide out of the atmosphere.

During launch the Mission Commander would take the left-hand couch where he could reach most of

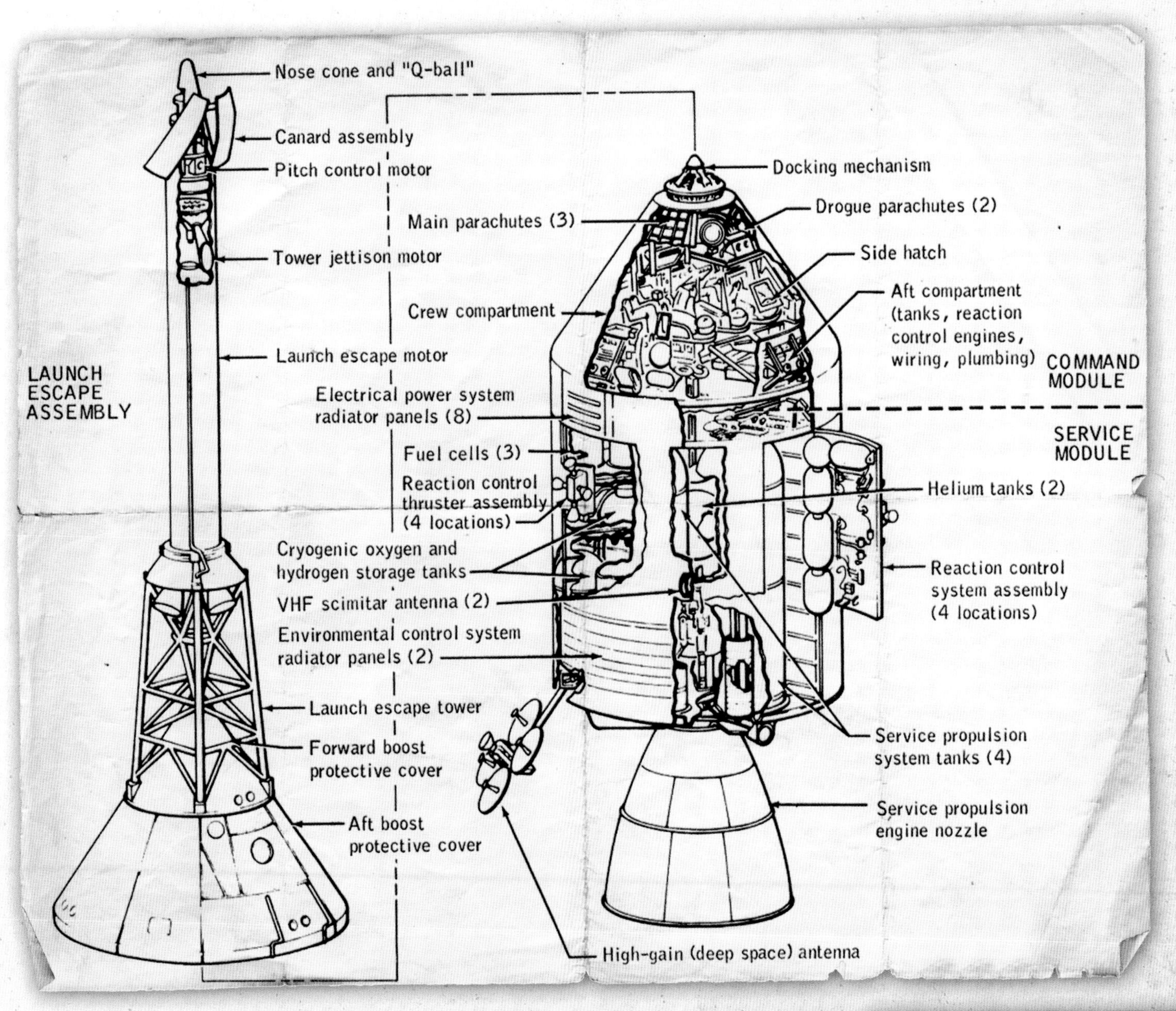


Diagram of the Apollo Command and Service Modules and Launch Escape Tower.

the flight instruments. The Command Module Pilot would take the center couch. The Lunar Module Pilot would occupy the right-hand couch, monitoring the spacecraft's systems. During reentry and other specific maneuvers, the Commander and Command Module Pilot would swap seats. Otherwise the couch would be folded away.

On the top of the CM was the docking mechanism, which mated with the Lunar Module (LM). Once docking was completed, it was removed to give access to the docking tunnel that allowed the astronauts to move between the two modules. Around the docking tunnel was the Earth Landing System—the drogues and main parachutes which slowed the craft after reentry and the pyrotechnic charges that deployed them.

## LAUNCH ESCAPE TOWER

Above the CM was the Launch Escape Tower (LET). It was about 33 feet (10 m) long and weighed about 8,000 pounds (3,629 kg). The lattice-work tower carried a cylindrical section that housed three solid-propellant rocket motors and a ballast compartment topped by a nose cone containing instruments.

It was activated automatically by the emergency detection system in the first 100 seconds or manually by the astronauts at any time from the launch pad to jettison altitude. The LET carried the CM to a sufficient height and to the side, away from the launch vehicle, so that it could land safely by parachute.

Extensive tests on the CM revealed that it was unlikely to survive a hard landing intact, so an extra rocket was

## The Reentry Fireball

When a spacecraft is launched, it needs protection from aerodynamic heating. This is caused both by the friction as the air passes over its surface and by compression of the air that cannot get out of the way fast enough. It is the same effect as the heating of a bicycle tire when you pump it up. As the air is compressed, the energy within a given volume increases.

The compression effect happens even more markedly on reentry. The speeds involved—and therefore the compression—is increased, resulting in temperatures of thousands of degrees, even approaching the temperatures found on the surface of the Sun. The advantage of reentry blunt-end first is that it creates the compression in front of the spacecraft, rather than around it.

The air gets so hot that the electrons are stripped from the atoms, creating a plasma. This forms a glowing fireball around the capsule, cutting any radio communication.

incorporated in the LET which would push the capsule out over the Atlantic in case it was activated on or near the launch pad.

With the Saturn V, the LET was jettisoned at about 295,000 feet (89,900 m), or about thirty seconds after ignition of the second stage, exposing the docking mechanism.

*It's a very sobering feeling to be up in space and realize that one's safety factor was determined by the lowest bidder on a government contract.*

Alan Shepard
American astronaut

Apollo 15 Command and Service Modules (CSM) in lunar orbit as photographed from the Lunar Module (LM) just after rendezvous.

## THE HEAT SHIELD

The base of the capsule was made from two shells separated by an insulating material. The inner shell was made from lightweight double-skinned aluminum and was an integral part of the capsule's pressure shell. The outer skin was a bonded fiberglass honeycomb filled with epoxy resin. As well as a wraparound heat shield, it also provided protection against micrometeorites.

The outer skin also accommodated ports for the discharge of waste water and urine, the communications antennas, and small rocket thrusters. The heat shield was thickest at the base where the heat of reentry would be the most intense. The heat shield was designed to char and burn away, carrying the heat with it in a process called ablation.

The rest of the capsule needed heat shields too. The heat on launch was absorbed principally through the boost protective cover, a fiberglass structure covered with cork which fitted over the CM like a glove. The boost protective cover weighed about 700 pounds and varied in thickness from about three-tenths of an inch (7.6 mm) to about seven-eighths of an inch (22.2 mm) at the top. The cork was covered with a reflective white coating. The cover was attached to the LET and they jettisoned together at around 295,000 feet (89,900 m) during a normal mission.

Once in space the CM still needed protection from the heat. The Sun raised the temperature of the side facing it to 280°F (138°C), while the side facing away from the Sun would remain at –280°F (–173°C).

During reentry the apex covering the nose of the cone would be jettisoned at around 24,000 feet (7,315 m) to expose the parachutes.

## THE GUIDANCE AND NAVIGATIONAL SYSTEM

The Guidance and Navigation and Control System was designed to keep Apollo on course, even when communication with Earth was interrupted. It had two major components:

1) The Inertial Measurement Unit, which used accelerometers, gyroscopes and, sometimes magnetometers to detect linear velocity and movement around the vehicle's three axes—pitch, roll, and yaw.

2) The Apollo Guidance Computer used the data to control the spacecraft. This had the power of a first-generation home computer. It could also be

programmed manually by the astronauts, who were provided with a sextant and telescope to make visual readings of the stars as backup.

## THE EARTH LANDING SYSTEM

As well as the drogues and parachutes needed to make a safe splashdown, the Command Module Pilot had some ability to steer the spacecraft. The craft's center of gravity would be tilted slightly during free fall, the tilt giving it a slight aerodynamic lift. Using the reaction control system—the small engines around the capsule that control pitch, roll, and yaw—it was possible to turn it to the left or right, or give it more lift to increase the range if the rescue zone was further away.

At around 27,000 feet (7,620 m) while still traveling at 320 mph (514 kph), a pyrotechnic charge would deploy the CM's first two drogue chutes. Then at 10,000 feet (3,000 m), with the CM still traveling at 160 mph (257 kph), the drogues would be jettisoned and pyrotechnic mortars would release the three main chutes. If at least two of the three worked, the capsule would splashdown at 20 mph (32 kph).

The orange and white main chutes each comprised half-an-acre of lightweight nylon fabric with two million stitches holding them together. They were attached to the CM by 1.5 miles (2.4 km) of suspension line. To get them into their compartments round the edge of the docking tunnel, they had to be compressed in a hydraulic press until they were the density of maple wood.

The Apollo 15 Command Module (CM) safely touches down in the mid-Pacific Ocean to conclude a highly successful lunar landing mission, 1971.

Splashdown tests were conducted in a giant pool behind manufacturer North American Aviation's factory at Downey, California. On one occasion the capsule bellyflopped, cracking the heat shield and sank. Ways had to be found to strengthen the outer skin without adding weight. There was also the danger of an unpredicted storm or high seas in the splashdown area, so inflatable bags were added to the capsule's collar to right it if it tipped over in the swell.

*The world is being Americanized and technologized to its limits, and that makes it dull for some people. Reaching the Moon restores the frontier and gives us the lands beyond.*

Isaac Asimov

## THE SERVICE MODULE

The Service Module (SM) was an unpressurized cylindrical structure, measuring 24 feet 7 inches (7.49 m) long and 12 feet 10 inches (3.91 m) in diameter. It remained attached to the CM by a fairing for the bulk of the flight and was only jettisoned before reentry. The fairing was half-an-inch thick (12.7 mm) and 22 inches (56 cm) high, comprising eight honeycomb aluminum panels and eight electrical power subsystem radiators which radiated excess heat into space from the power cells below.

Below the SM were six radial compartments around a center section 44 inches (112 cm) in diameter. This contained two helium tanks and, below them, the Service Propulsion System (SPS) engine which slowed the spacecraft to enter lunar orbit, provided the power to propel it back to Earth, and made course corrections along the way. Its nozzle extension skirt protruded more than 9 feet (3 m) below the aft bulkhead of the module which was protected from its blast by a heat shield.

As there was no backup for the SPS engine, it had to work first time at critical moments during the mission. This was ensured by using hypergolic fuel. When the dimethyl-hydrazine propellant and the nitrogen tetroxide oxidizer were mixed they spontaneously combusted, eliminating the need for an ignition system. The need for pumps—with their possibly unreliable moving parts—was also eliminated by pressurizing the tanks with helium which is inert.

The six radial sectors varied in size—1 and 4 were 50°, 3 and 6 were 60°, and 2 and 5 were 70°. Sector 1 was left empty until later flights when it contained the Scientific Instrument Module. Sector 2 contained the SPS engine's oxidizer sump tank, which was kept refilled from the oxidizer storage tank in Sector 3.

Sector 4 carried the fuel cells for the electrical power system, along with the hydrogen and oxygen needed to run them. Fuel for the SPS engine was kept in a sump tank in Sector 5. This was topped by the fuel storage tank in Sector 6.

### The Fuel Cells

A reliable source of electrical power was vital to any space mission. Earlier Mercury and Gemini missions had depended on batteries, but they would be too heavy if they were to supply the power for a two-week mission. Unmanned flights into the inner solar system had used solar power. But the panels were unwieldy and as the engines had to be fired several times during an Apollo mission they would have been subjected to mechanical stress that risked damaging them.

The answer was fuel cells. These supply energy over a much longer time than a battery as they are continuously supplied with fuel and oxygen. In Apollo's fuel cells developed by Pratt & Whitney Aircraft Division, hydrogen and oxygen reacted together to produce one kilowatt of electricity per cell. A useful byproduct was water that was passed into the Command Module by an umbilical connection. It could be used for cooling the electrical system as well as drinking, washing, and rehydrating food.

The Command Module also carried five silver-oxide-zinc batteries as an extra power source when extra electricity was required and during reentry when the Service Module was jettisoned.

# CHAPTER 10

# THE LM

Once it was decided to use Lunar Orbit Rendezvous (LOR) to put a man on the Moon, a lunar lander had to be built. Also known as the Lunar Excursion Module, the Lunar Module (LM) was the portion of the Apollo spacecraft that would actually land on the Moon and blast off back into orbit again. It would carry a crew of two. By 1962, manufacturer Grumman had spent two years studying the possibilities of building the flimsy craft. The company's expertise was building World War II fighter planes to land on aircraft carriers, but the managers were keen to get involved with the Apollo program.

## DESIGNING THE LUNAR MODULE

LOR meant that the landing craft could be built in two stages, each with its own engine. The engine in the descent stage could slow the lander so that it dropped out of lunar orbit, then slowed the craft further as it approached the surface so that it could make a soft landing. It would also incorporate landing gear.

A smaller ascent stage would then blast off, using the bulky descent stage as a launch pad. It would carry the crew back into orbit to rendezvous and dock with the Command/Service Module (CSM). This configuration saved weight and won Grumman the contract. It was signed in January 1963, two years after work on the CSM had started.

The brief was clear. The LM would have to carry two men and 250 pounds (113 kg) of equipment to the surface of the Moon, keep them alive there for forty-eight hours, then return them to the CSM. To do that, the lander would have to make a controlled descent to the surface. This would require a new kind of throttle-controlled jet engine.

The life-support systems onboard would have to be every bit as reliable as those on the CSM, but much lighter. What's more, there would be no opportunity for a test flight on Earth. When it reached the Moon, everything would have to work perfectly.

*We are banking our whole program on a fellow not making a mistake on his first landing.*

Peter Conrad, Apollo 12 Mission Commander

## HOW TO LOSE WEIGHT FAST

NASA instructed Grumman that, to save weight, everything superfluous must be discarded. The LM immediately lost one of the five legs it had in the original design. The original circular chassis was swapped for a square one with a leg at each corner, though ultimately an octagonal base was used.

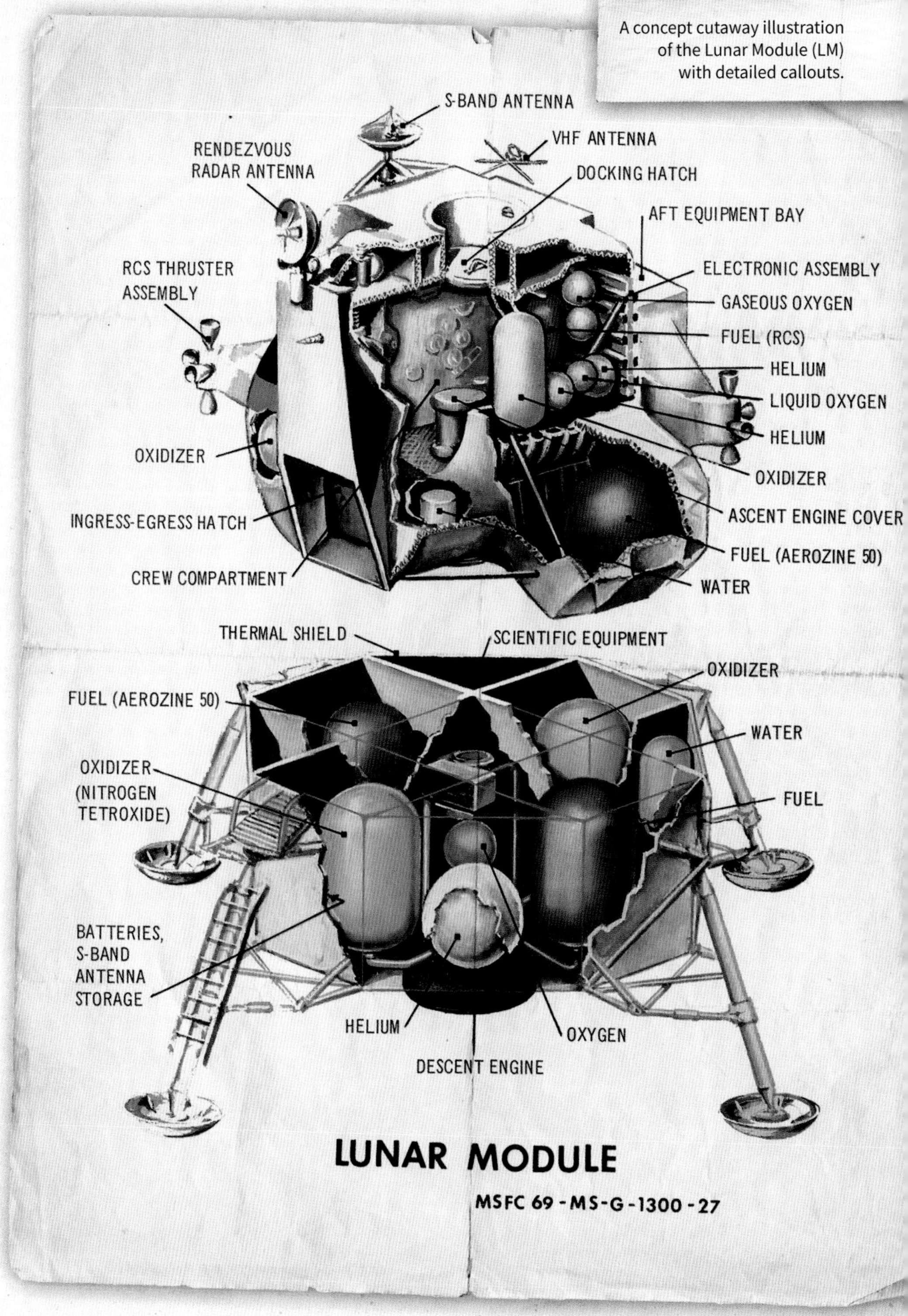

A concept cutaway illustration of the Lunar Module (LM) with detailed callouts.

The engine would be in the middle of the descent stage with four propellant tanks around it, instead of six, saving the weight of additional piping. The engine gimbaled to provide steering and to compensate for adjustments in the center of gravity as the propellant was used up.

Size was an obvious constraint. At 23 feet (7 m) high and 14 feet (4.26 m) wide, on launch the LM would have to sit under the CSM with its legs folded in an aluminum sleeve on top of the third stage of the Saturn rocket known as the Spacecraft Lunar Module Adapter (SLA).

## LANDING THE LM

Once the preparatory orbits of the Earth had been completed and the trans-lunar injection had begun, the CSM separated and the sleeve opened like the petals of a flower. The CSM then turned round and came back to dock with the LM. The LM pilot floated through the docking hatch to power up the flimsy little craft.

Once in lunar orbit, the pilot was joined by the commander and the docking hatch closed. The LM's legs then unfolded and it separated from the CSM. The LM's engine fired to slow the craft and it descended to make a powered landing. When leaving the Moon, the LM separated from the descent section, leaving it on the surface, while the ascent engine put it back in orbit where it would dock with the CSM. After the crew had transferred into the CSM, the ascent section of the LM would be abandoned.

When the Apollo 11's *Eagle* made the first landing on the Moon, it carried 18,000 pounds (8,165 kg) of fuel to land and 5,200 pounds (2,359 kg) to carry it back to the orbiting CSM, with an extra 600 pounds (272 kg) for maneuvering. It carried eighteen rockets and legs were extended by explosives. There were eight types of radio onboard and two kinds of radar, connected by thirty miles of wiring, and at each corner there were peroxide thrusters.

# TRANSFER TO LM

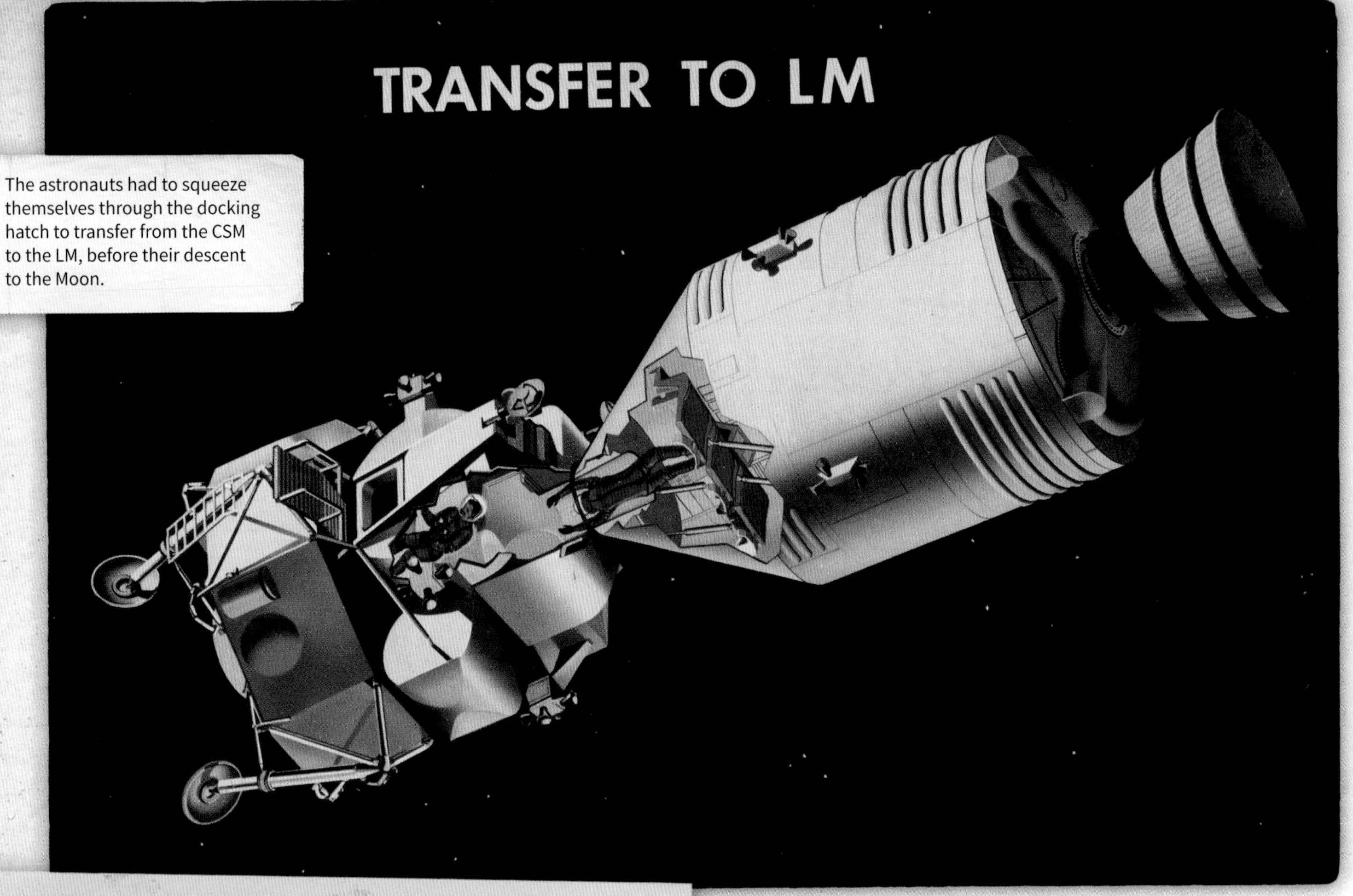

The astronauts had to squeeze themselves through the docking hatch to transfer from the CSM to the LM, before their descent to the Moon.

Neil Armstrong inside the LM cabin during simulation training.

## INSIDE THE CABIN

The design of the original cabin also had to be scrapped as it soon became clear that there was no room for the astronauts' bulky EVA suits with the backpacks and there would have to be somewhere to stow the rock samples they had collected.

More weight was saved by throwing out any seating. The astronauts would have to stand during the fifteen-minute descent, tethered by the waist and with their feet Velcroed to the floor. Their legs would act as shock absorbers during landing and liftoff, and fold-down arm rests would be added so they could steady themselves.

With the pilot standing, the cockpit was redesigned with smaller, triangular windows tilted downward at head height, rather than the helicopter dome originally envisaged. These bulged outward from the air pressure. Another rectangular window was added in the roof to give the Commander a view of the CSM during docking.

The astronauts had to get in and out of the cabin once on the surface, so a ladder was added to one of the legs. A small porch was added so that an astronaut in a bulky spacesuit could reach the top of the ladder. The craft ended up looking like an insect from outer space and many in NASA and Grumman nicknamed it the "Bug."

## THE DESCENT STAGE

The octagonal base had a span of 14 feet 1 inch (4.3 m). The interior was divided by walls, giving a cross of square compartments surrounding the tanks of propellant and oxidizer, with the engine occupying the center square. The retractable legs hinged out from the ends, doubling the span to 31 feet (9.4 m). They had footpads to prevent them sinking into the moondust. Beneath three of them were 68-inch (172 cm) probes. When they touched the surface, a blue "contact" light lit inside the cabin telling the pilot to cut the engine and drop the last few feet to the ground.

This left four triangular-shaped compartments between them, holding water, helium to pressurize the tanks, scientific equipment, TV cameras, tools and, on later missions, the lunar rover. The whole thing was covered in sixteen layers of gold-colored Mylar aluminized foil tape that protected against micrometeorites and the extremes of temperature, without adding much weight.

## THE VARIABLE THRUST ENGINE

Time was short so Grumman contracted out the design of the innovative variable thrust engine that would be needed to make a soft landing. Rocketdyne built a prototype that varied the thrust by adding inert helium to the propellant. But the design that was picked came from Space Technology Laboratories. This used a simple mechanical device like a car's throttle that varied the amount of fuel going into the engine. At maximum thrust, it could lift 5 tons (4,535 kg). Throttle back and it could gently lift a newborn baby.

The walls of the propellant tanks were so thin that, it was said, they bulged when filled on Earth. The Aerozine 50 and the nitrogen tetroxide oxidizer were so corrosive that the engine only had a forty-day life after it had been exposed. That meant that they could not be tested before the rocket was assembled. The first test would be performed 240,000 miles (386,242 km) from home.

The LM's final descent to the lunar surface. Long before the days of computer-generated images, all NASA's concepts had to be hand-drawn by highly-skilled illustrators.

## THE ASCENT SECTION

Like the SPS engine, the ascent engine had to be 100 percent reliable. If it failed, the astronauts would be stranded on the Moon with no possibility of rescue. Again it used hypergolic reactants forced into the engine bell by pressurized helium. The tanks were mounted on either side of the cabin.

The makers, Bell Aerosystems, further simplified the engine by replacing the conventional piped cooling system with ablative coating that charred and burned off. But NASA was concerned about combustion instability—a problem that had plagued the Saturn's F-1 engines. Grumman put the Bell engine through the same strictures by setting off explosives in the engine while it was firing. Under certain conditions, they could produce dangerous oscillations that did not damp down. For two years, Bell tried everything they could to fix it.

In 1967, with months to go before the LM's first flight, NASA called in Rocketdyne. They came up with a new injector system. By June 1967, the engine had survived fifty-three explosive tests with the oscillations damped down in under 400 milliseconds and the LM was ready for delivery to Cape Kennedy.

## THE CREW COMPARTMENT

The crew compartment was shaped like a tube 11 feet 10 inches (3.6 m) long and 16 feet (4.87 m) in diameter. This was mounted horizontally across the top of the ascent engine. It gave the two astronauts enough room to stand side by side in their pressure suits, or lie on the floor if they needed some sleep, though from Apollo 12 onward, hammocks were provided. These were extremely comfortable in one-sixth gravity.

During landing and liftoff, the crew stood at the front, surrounded by the main controls. Behind them was the equipment bay which contained the communications equipment and life support systems. There were also storage bays for their EVA suits, food, and other necessities.

Access to the crew compartment from the CM was through the docking hatch in the roof. The second hatch used to exit and enter the lander once it was on the Moon was in the middle of the control panel, between the two crew stations. Looking from the inside, the hinge was on the right and the hatch door opened outward, so the Commander, who stood on the left, would leave the lander first.

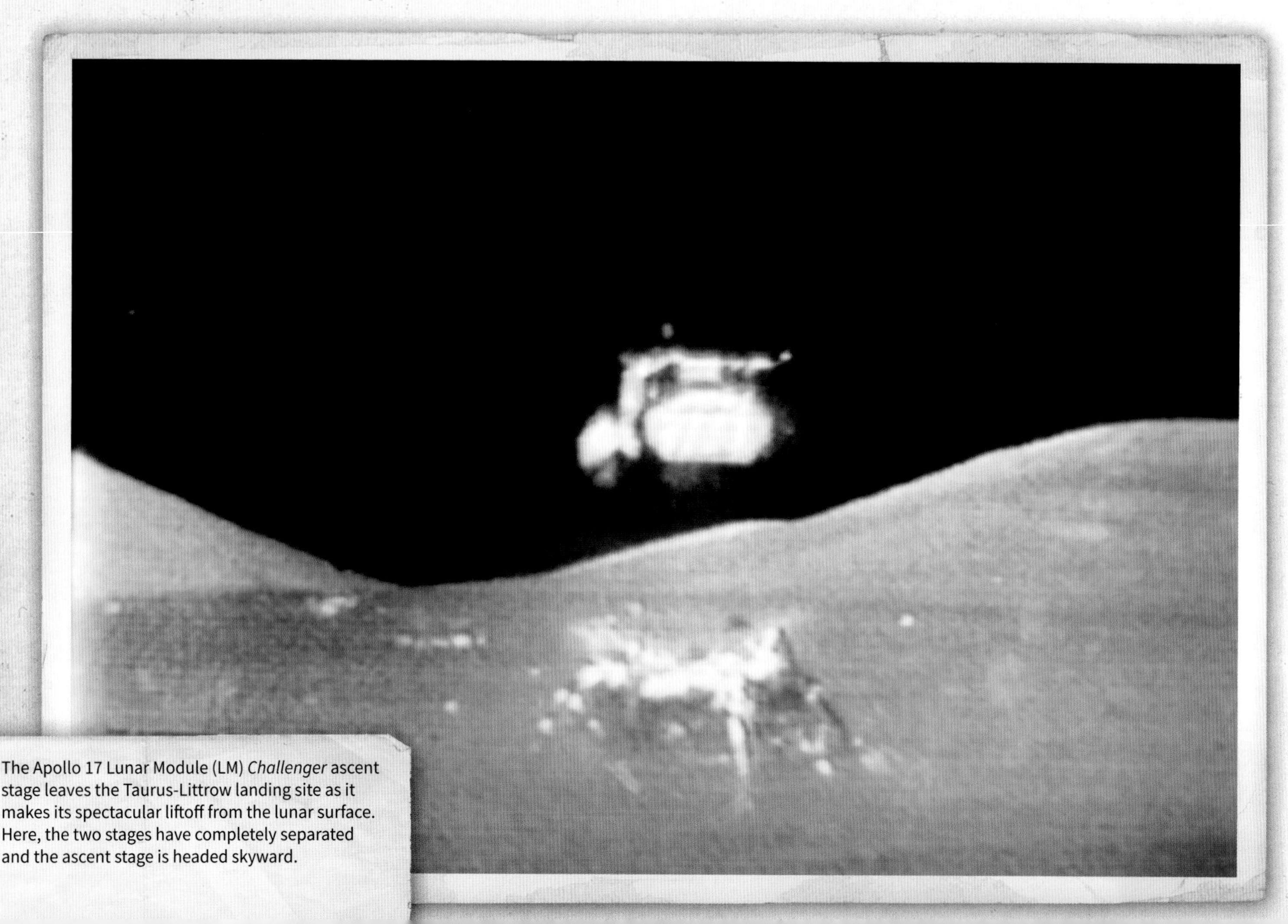

The Apollo 17 Lunar Module (LM) *Challenger* ascent stage leaves the Taurus-Littrow landing site as it makes its spectacular liftoff from the lunar surface. Here, the two stages have completely separated and the ascent stage is headed skyward.

To get out, the astronauts would have to be on all fours and crawl out backward across the porch, then step down the nine rungs of the ladder.

The skin of the cabin was just 0.012 inches (0.3 mm) thick. This is about the same as three layers of kitchen foil. It buckled under the g-force when the Saturn V launched, but was held rigid when pressurized like a balloon. As in the CM, weight was saved by using pure oxygen as it was comfortably breathable at one-third the pressure of air.

During landing and launch, the crew plugged their pressure suits into the LM's life-support system for oxygen and cooling water. On later missions, they plugged in to top-up their backpacks.

---

*A human being is the best computer available to place in a spacecraft ... It is also the only one that can be mass produced with unskilled labor.*

Wernher von Braun

---

## FLYING THE LM

Four clusters of small rocket thrusters were mounted on the four corners of the LM, allowing it to maneuver accurately for landing, rendezvous, and docking. These used the same hypergolic fuel as the main engines, which was stored on tanks around the outside of the cabin.

The problem was that the LM's flight characteristics were markedly different when it was descending than when it was docking, having jettisoned the descent stage and used up most of the fuel. This meant it had to be controlled by a flight computer called the Primary Navigation and Guidance System (PNGS), also known as "Pings." This compensated for the changing flight characteristics, interpreting the movements of the control stick. It was, essentially, the first "fly by wire" system.

## THE DANGERS OF WEIGHT GAIN

Originally the LM was supposed to have weighed just 22,000 pounds—11 tons or 10,000 kg. That proved impossible to achieve and the weight limit gradually increased. By the beginning of 1965, it was 32,500 pounds—over 16 tons or 14,740 kg. Even this was proving hard to attain.

There were problems with the fuel cells so the designers reverted to the heavier silver-zinc batteries. There were also questions over reliability of the PNGS guidance system, so a new AGS, or Abort Guidance System, was added which would take the craft back into orbit if there was a problem.

But it was vital to keep the weight down. Every extra pound meant three more pounds of propellant had to be carried. The ascent would also take longer, exposing the crew to greater risks. And the heavier the spacecraft the more the Saturn V would have to lift.

In July 1965, NASA took the simple expedient of offering Grumman a bonus of $25,000 for every pound they saved. Engineers pored over the designs to see what weight savings could be made. The most striking was replacing the descent-stage thermal shields with gold Mylar tape.

But it was too late to make any major design changes, so Grumman instigated Operation Scrape. Metal was shaved from individual bolts. The thin varnish that protected the aluminum frame from the weather was stripped off and acid was used to etch surface components. Eventually another 2,500 pounds (1,134 kg) was saved, bringing the LM in under the weight limit and netting Grumman a bonus of $62.5 million.

### Aerozine 50

Aerozine 50 is a 50:50 mix by weight of hydrazine and unsymmetrical dimethylhydrazine, originally developed in the late 1950s by Aerojet General Corporation as a hypergolic fuel for the Titan II ICBM rocket engine. It was so toxic that, to test it, Grumman had to build a new facility at White Sands, New Mexico. Similarly, the oxidizer nitrogen tetroxide was so lethal that any more than five parts per million would eat away at your lungs. When it leaked it formed a poisonous red cloud in the sky around White Sands. The police had to evacuate the area until it dissipated.

PART FOUR

# THE APOLLO MISSIONS

*FLY ME TO THE MOON*
*LET ME PLAY AMONG THE STARS*
*LET ME SEE WHAT SPRING IS LIKE*
*ON JUPITER AND MARS.*

"FLY ME TO THE MOON" (1954)
LYRICS BY BART HOWARD

# CHAPTER 11

# WE MUST BE BOLD

The Apollo 1 disaster led to a thoroughgoing redesign of the entire Apollo spacecraft over eighteen months while all flights were suspended. Fireproof fabrics and paint were used. The redesign was tested on an unmanned orbital flight using a Saturn 1B. Work also had to be done on the Saturn V rocket.

## THE SPACE RACE HEATS UP

Meanwhile on January 3, 1966, the Soviet Union landed an unmanned automatic lunar station on the Moon in the Ocean of Storms. It sent back photographs from the Moon and had a pressurized cabin suitable for human beings to survive in.

However, Sergei Korolev died after an operation for colon cancer. He was replaced by his deputy Vasily Mishin who had a series of failures to his name. Four rockets exploded on launch.

Then 40-year-old Vladimir Komarov died onboard Soyuz 1 when the spacecraft's drogue and main breaking parachute failed to deploy properly and the capsule crashed into the ground on April 24, 1967. Yuri Gagarin died the following year in a plane crash. Mishin was eventually succeeded by Valentin Glushko in 1974 when the Kremlin decided to consolidate the entire Soviet space program into one organization and gave up on going to the Moon.

*If we shall send to the Moon, 240,000 miles away ... a giant rocket ... made of new metal alloys, some of which have not yet been invented ... fitted together with a precision better than the finest watch—then we must be bold.*

President John F. Kennedy
September 12, 1962

But the space race was not yet won. By 1967, the deadline given by President Kennedy for a successful Moon landing was looming and the program was running disastrously behind schedule. The redesign of the CM was not going well and the LM, built by Grumman, did not meet the specifications.

## APOLLO 4

The first fully assembled Saturn V was finally carried slowly out to Launch Complex 39 at NASA's Kennedy Space Center on August 26, 1967. Two months of testing followed. Twice the kerosene fuel and liquid oxygen had to be pumped out. Then on November 6, once again twenty-eight truckloads of liquid hydrogen and ninety truckloads of liquid oxygen were pumped into the tanks. The kerosene was brought in on twenty-seven separate rail cars. Apollo 4 was ready for launching. The unmanned mission was the first "all-up" test of the three stages of the Saturn V rocket and was designed to test all aspects of the launch vehicle.

Nine seconds before the launch at 7:00 a.m. EST on November 9, 1967, the kerosene and liquid oxygen began rattling down the pipes toward the first-stage engines. The central engine was ignited first, then the pairs in opposing corners at intervals of 300 milliseconds. Ice that had formed on the chilled tanks of the upper stages began to come loose. At 160 million horsepower, the rocket strained at the huge locks holding it down.

Once the Instrumentation Unit sensed that the F-1 engines had achieved their maximum thrust, it released the locks and Apollo 4 began to lift off. The shock wave rocked the VAB and the press and VIP stands 4 miles (6.4 km) away. At liftoff, the vibration from the Saturn V showered technicians with dust and debris from the ceiling of the Launch Control Center. Veteran CBS news anchor Walter Cronkite who was the TV commentator at the time captured the excitement in his inimitable style:

> *The building's shaking! This big blast window is shaking! We're holding it with our hands! Look at that rocket go into the clouds at three thousand feet! … you can see it … you can see it … oh the roar is terrific! …*

## CLEARING THE TOWER

It took twelve seconds to clear the tower, yawing away for safe clearance as it passed the top. Then it pitched and rolled onto the correct course, with the four outer engines gimbaling outward to give stability.

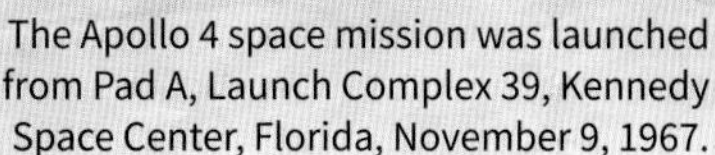

The Apollo 4 space mission was launched from Pad A, Launch Complex 39, Kennedy Space Center, Florida, November 9, 1967.

At 6,500 feet (1,981 m) Apollo 4 was traveling at over 1,100 mph (1,770 kph). The five F-1 engines in the first stage continued to burn until the giant rocket was 38 miles (61 km) high. Then they cut out. Separation occurred 600 milliseconds later. Eight solid rockets fired, pushing the first stage away. After thirty seconds, the inter-stage ring fell away, along with the LET on top of the CM.

Two of the second-stage engines gave out during liftoff, causing the rocket to keel over. However, its guidance system gimbaled the engines and righted the vehicle. The stage-two engines burned for 6 minutes 30 seconds, lifting the craft to an altitude of 108 miles (174 km) and a speed of 17,400 mph (28,000 kph). The Instrumentation Unit then sensed the tanks were empty and jettisoned stage two, leaving the stage three S-IVB to carry the CSM into a near circular orbit at an altitude of 115 miles (185 km)—the height of the initial parking orbit on the lunar missions.

### REIGNITION AND ATTITUDE THRUSTS

After two orbits, the S-IVB's first reignition put the spacecraft into an elliptical orbit with a high point of 11,200 miles (18,000 km). It remained at that altitude for 4 hours 30 minutes being soaked in solar radiation, while engineers in Mission Control monitored the cabin environment until they were satisfied that the conditions onboard could safely support life.

Then the attitude thrusts upended the spacecraft, tipping the nose down toward the Earth. The Service Module's service propulsion system engine then fired to increase reentry speed to about 24,900 mph (40,000 kph), simulating a return from the Moon. The Service Module was jettisoned soon after. The capsule was traveling at seven miles a second when it hit the outer atmosphere, the fastest reentry ever attempted.

Heated to over 9,000°F (5,000°C), the capsule's underbelly glowed white hot. The wave of compressed air ahead of it turned to a plasma and cameras pointing out the windows of the capsule recorded ribbons of orange and white streaking passed. Next the capsule's heat shield began to disintegrate, as it was designed to do, and fragments streamed passed the window.

After 8 hours 36 minutes of flight, the CM splashed down in the Pacific Ocean, about 10 miles (16 km) from the aircraft carrier USS *Bennington*, the prime recovery ship. The Saturn V's first outing had been a success and NASA's confidence soared.

## APOLLO 5

The unmanned Apollo 5 was the first LM test flight. LM-1 had arrived at the Cape onboard Aero Spacelines Super Guppy on June 23, 1967, after problems with the ascent engine had been overcome. It was supposed to have flown in Apollo 4. However, at the Kennedy Space Center it was taken into a clear room where NASA inspectors crawled all over it.

They found hundreds of technical faults. There were broken wires and leaks in the pressurized propellant system. On a mission such as this, even the tiniest leak of highly flammable hypergolic fuel could prove disastrous. LM-1 was grounded until every one of the faults was fixed. They ended up inspecting the joints with X-rays before NASA were satisfied.

## What is a Gimbal?

A gimbal is a platform that can pivot. Instead of being fixed to an unmoving base, an object on a gimbal can rotate along at least one axis. In the world of aeronautics, an aircraft in flight is free to rotate around three axes called roll, pitch, and yaw.

To understand roll, pitch, and yaw, you need to visualize three lines running through an airplane and intersecting at right angles at the airplane's center of gravity. First imagine a line that runs through the front of an aircraft and out the back. A rotation along this nose to tail axis results in a roll.

Imagine another line running through both wings of the plane. A rotation along this side-to-side axis is a change in pitch. The plane either climbs or dives, depending on the direction of the pitch. A full circle would be a loop-the-loop.

Finally, think of an imaginary vertical line that comes out of the top and bottom of the plane. This is the yaw axis. Rotating along this vertical axis results in a change in direction for the plane, either right or left.

An object mounted on three or more gimbals can turn in nearly any direction. The device has been known since antiquity, and was first described in 3 BC by Philo of Byzantium, although it has no single identifiable inventor. The first aircraft to demonstrate active control about all three axes was the Wright brothers' 1902 glider.

After four months of tests and repair LM-1 was mated to the launch vehicle on November 19. However, on December 17, 1967, another LM under test failed in the Grumman Aircraft Engineering Corporation ascent-stage manufacturing plant. A window in LM-5—which would become Apollo 11's LM *Eagle*—shattered during its initial cabin pressurization test. The following day, it was decided to replace the window with aluminum plates as a precaution.

After a long delay, the liftoff was scheduled for January 22, 1968, and the launch vehicle was to be a Saturn 1B, originally designated for Apollo 1. It was undamaged in the accident.

## ORBITAL PROPULSION TESTS

After a perfect launch, the S-IVB second stage ignited to insert the spacecraft into an Earth orbit 101 by 138 miles (163 by 222 km). The nose cone was jettisoned and after a coast of 43 minutes 52 seconds the LM was separated. The LM entered an orbit 103 by 138 miles (167 by 222 km). It had no legs because it was not going to land anywhere, but the integrity of the cabin could be tested along with the attitude thrusters, the throttling capabilities of the descent section, the stage separation, and the ascent engine.

After two orbits, the descent propulsion system (DPS) was ignited. A thirty-nine second burn was planned, but it was cut short after only four seconds. The burn was designed to simulate deceleration for descent to the lunar surface, but was stopped prematurely due to overly conservative programming of the flight software. An alternate flight plan was put into effect, in which the DPS fired for twenty-six seconds at ten percent thrust and then for seven seconds at maximum thrust.

A third DPS firing was performed thirty-two seconds later, consisting of a twenty-six second burn at ten percent thrust and two seconds at maximum thrust. This was followed by a burn to simulate an abort during the landing phase, where the ascent propulsion system (APS) was ignited simultaneously with the DPS being shut down. The APS burn lasted sixty seconds, followed by a firing of 6 minutes 23 seconds during which the APS fuel was depleted.

At the end of the 11 hour 10 minute test period, both LM stages were left in orbit to eventually reenter the Earth's atmosphere and burn up somewhere over the west coast of Panama. Despite the initial premature DPS shutdown, the mission was deemed a success. The next LM to fly in space would carry its first crew.

The Lunar Module-1 being moved into position for mating with Spacecraft Lunar Module Adapter (SLA)-7 in the Kennedy Space Center's Manned Spacecraft Operations Building. LM-1 and SLA-7 were flown on the Apollo 5 unmanned space mission.

## APOLLO 6

Apollo 6 was the second "all-up" unmanned test of the Saturn V. The plan was for the first three stages to place the spacecraft into low Earth orbit with the third stage still attached. To simulate a translunar injection sending the CSM to the Moon, the third stage would reignite to send the spacecraft into a highly elliptical Earth orbit.

Soon after, the Service Module's engine would fire to slow the spacecraft down in a simulation of a direct-return abort when the spacecraft is returned directly to Earth. It was planned for Apollo 6 to reach an altitude of about 14,000 miles (22,530 km) before descending back toward Earth. The SM engine would then fire again to increase the spacecraft's speed on reentry to about 25,000 mph (40,234 kph) to mimic a lunar return as it had done on Apollo 4.

The second Saturn V lifted off from the Kennedy Space Center's Launch Complex 39A on April 4, 1968, but the mission ran into trouble from the start. Two minutes into the flight, the first stage experienced about thirty seconds of vertical oscillations known as "pogo effect" which caused no serious damage but would have been very uncomfortable for any crew.

Then during the second stage burn, two of the five engines shut down prematurely. The remaining three engines burned longer to compensate for the reduced thrust. The third stage was also supposed to burn longer to propel Apollo 6 into orbit, but when it was time for the third stage to restart, it would not reignite.

The flight control team decided on an alternate mission plan. They would separate the spacecraft from the Saturn V's third stage and use the SM engine instead to reach the planned altitude of nearly 14,000 miles (22,530 km). But this used so much fuel that there was not enough left for its second burn, so reentry occurred at less than the planned speed. Apollo 6 splashed down in the Pacific Ocean after a troubled flight of 9 hours 57 minutes and was recovered by the aircraft carrier USS *Okinawa*.

However, when the camera pod jettisoned from the Apollo 6 second stage was recovered, NASA scientists found that it had captured spectacular images of the first-stage separation.

But the failure of Apollo 6 was overshadowed by other events. Four days before the flight President Lyndon B. Johnson announced that he would not seek reelection—and an hour after the splashdown, Martin Luther King was assassinated in Memphis, Tennessee.

## Aero Spacelines Super Guppy

As transporting parts by barge was slow and expensive, NASA shifted the LM-1, the Apollo 11 Command Module, and other large pieces of hardware for the Apollo program on the outsize cargo freight aircraft known as the Super Guppy. The wide-bodied cargo plane used the fuselage of the C-97J Turbo Stratocruiser, the military version of the 1950s Boeing 377 Stratocruiser passenger plane. The fuselage was lengthened to 141 feet (43 m), making the length of the cargo compartment 94 feet 6 inches (28.8 m). This was ballooned out to a maximum inside diameter of 25 feet (7.6 m). The entire nose section hinged, giving access to the full width of the cargo hold. It could carry a load of 54,000 pounds (24,494 kg) and cruise at 300 mph (480 kph).

The Apollo 11 spacecraft Command Module is loaded aboard a Super Guppy Aircraft at Ellington Air Force Base for shipment to the North American Rockwell Corporation at Downey, California.

One of the images captured by the camera onboard Apollo 6 showing the interstage section of the Saturn V rocket falling away.

CHAPTER 12

# OUT OF THIS WORLD

## APOLLO 7

Apollo 7 was to be the first manned mission. Onboard were Commander Walter Schirra, Command Module Pilot Donn Eisele, and Lunar Module Pilot Walter Cunningham. The launch vehicle was a Saturn 1B and blastoff was at 11:02 a.m. EST on October 11, 1968.

The primary objectives for Apollo 7 were simple: "Demonstrate CSM/crew performance; demonstrate crew/space vehicle mission support facilities performance during a manned CSM mission; demonstrate CSM rendezvous capability." It was thought these objectives could be met within three days but the mission would be open-ended up to eleven days "to acquire additional data and evaluate the aspects of long duration manned spaceflight." These extra days did leave plenty of time for "sight-seeing" and taking pictures of weather and terrain that had never been seen before.

### RIDING LIKE A DREAM

It was a hot day at Cape Kennedy, but the heat was tempered by a pleasant breeze when Apollo 7 lifted off, further heating the air with two tongues of orange-colored flame. The Saturn 1B, in its first trial with men aboard, provided a perfect launch, and its first stage separated 2 minutes 25 seconds later. The S-IVB second stage took over, giving the astronauts their first ride on top of a tank of liquid hydrogen.

Walter Schirra, who had already flown with Mercury and Gemini, reported 5 minutes 54 seconds into the mission: "She is riding like a dream." Riding a Titan had been much bumpier.

About five minutes later, an elliptical orbit was achieved 140 by 183 miles (225 by 294 km) above Earth. The S-IVB stayed with the CSM for about one-and-a-half orbits, then separated. Schirra fired the CSM's small rockets to pull 50 feet (15 m) ahead of the S-IVB, then he turned the spacecraft around to perform the docking necessary to extract the LM from the Spacecraft Lunar Module Adaptor (SLA), the sleeve that housed the LM on top of the S-IVB, for a Moon landing. However, this maneuver ran into difficulties.

The next day, when the CSM and the S-IVB were about 80 miles (130 km) apart, they attempted to rendezvous with the tumbling 59-feet (18-m) second stage. Approaching within 70 feet (21 m), they reported that the SLA's four hinged aluminum panels which opened outward to reveal the LM had not fully deployed, preventing docking. If that happened on a Moon mission, it would have proved impossible to extract the LM without damage and the mission would have had to be aborted. In future flights, the panels would be jettisoned explosively.

Nevertheless, the SPS engine performed faultlessly. This was crucial as there was no redundant or backup system to switch to. At crucial times, the engine simply had to work or they would not get back home. On Apollo 7, there were eight nearly perfect firings out of eight attempts.

## YABA-DABA-DOO!

On the first, the crew had a real surprise. In contrast to the smooth liftoff of the Saturn, the blast from the SM engine jolted the astronauts. Schirra morphed into Fred Flintstone yelling "Yaba-daba-doo" at the top of his voice. Donn Eisele added more graphically it was a real boot in the rear as they were plastered back into their seats. But the engine did what it was supposed to do each time it fired.

The CSM performed perfectly for the 10.8 days they stayed in space. This was longer than they needed to travel to the Moon and back. With few exceptions, the systems in the spacecraft operated as they should. Occasionally, one of the three fuel cells supplying electricity to the craft developed some unwanted high temperatures, but load-sharing hook-ups among the cells prevented any power shortage.

The crew complained about noisy fans in the environmental circuits and turned one of them off. That did not help much, so the men switched off the other. The cabin stayed comfortable, although the coolant lines sweated and water collected in little puddles on the deck, but the crew simply vacuumed the excess water out into space with the urine dump hose.

A momentary shudder went through the Mission Control Center in Houston when both AC buses dropped out of the spacecraft's electrical system. This

Apollo 7, the first manned Apollo space mission, is launched from the Kennedy Space Center, October 11, 1968.

coincided with the automatic cycles of the cryogenic oxygen tank fans and heaters. The problem was solved by manual resetting of the AC bus breakers.

## VISION OF THE HEAVENS

Three of the five spacecraft windows fogged because of improperly cured sealant compound, a condition that would not be fixed until Apollo 9. Visibility from the spacecraft windows ranged from poor to good during the mission. Shortly after the LET jettisoned, two of the windows had soot deposits and two others had water condensation. Two days later, however, Cunningham reported that most of the windows were in fairly good shape, although moisture was collecting between the inner panes of one window. On the seventh day, Schirra described essentially the same conditions.

Even with these impediments, the visibility was adequate. The windows used for observations during rendezvous with the S-IVB remained almost clear. However, navigational sighting with a telescope and a sextant on any of the thirty-seven preselected stars was difficult if done too soon after a waste-water dump.

Sometimes they had to wait several minutes for the frozen particles to disperse. Eisele reported that unless he could see at least forty or fifty stars at a time he found it hard to decide which part of the sky he was looking toward. On the whole, however, the windows were satisfactory for general and landmark observations and for out-the-window photography of the heavens.

## WASTE MANAGEMENT

Despite minor irritations, such as smudging windows and puddling water, most components supporting the well-being of the spacecraft and crew worked as planned. The waste-management system for collecting

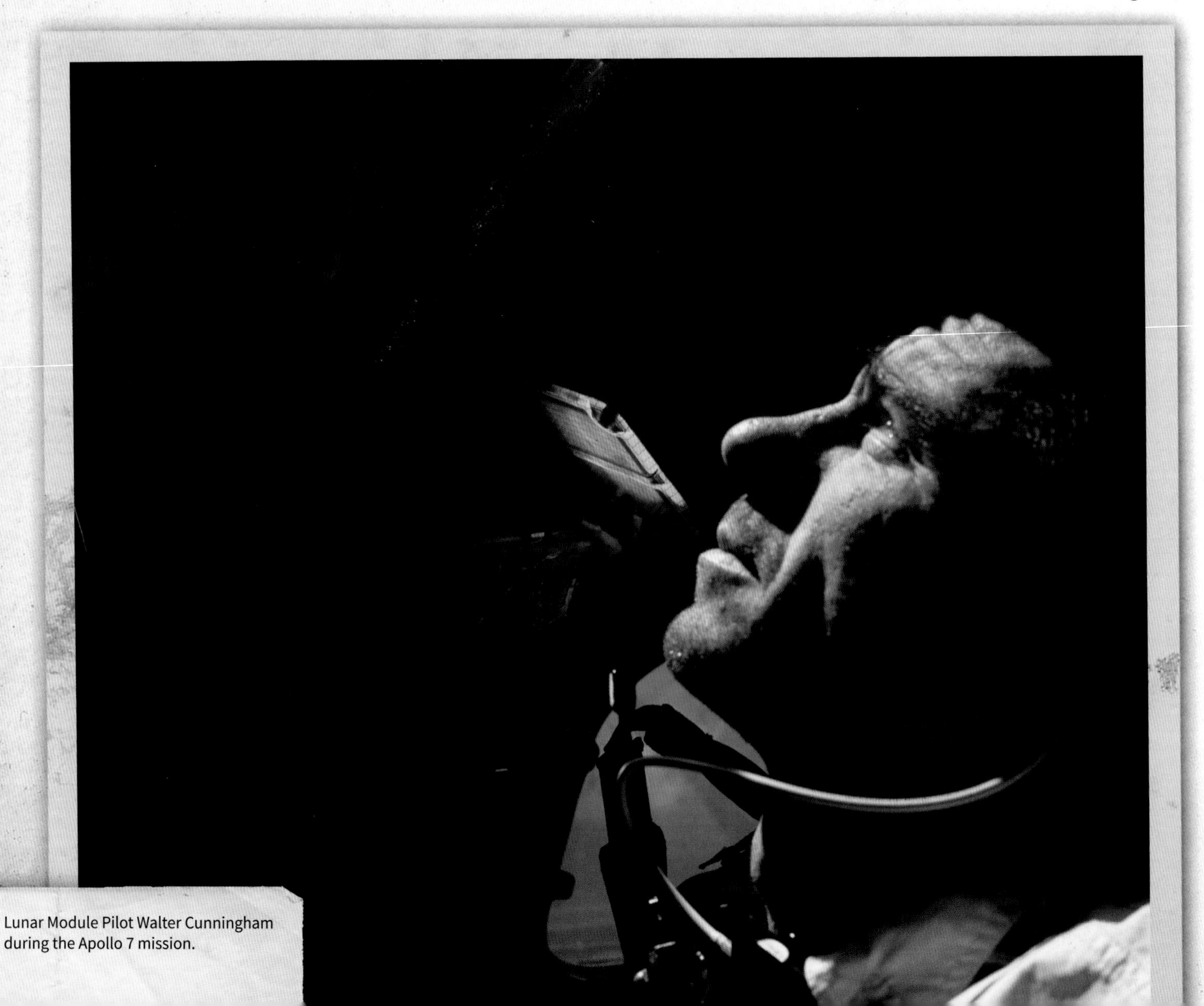

Lunar Module Pilot Walter Cunningham during the Apollo 7 mission.

solid body wastes was adequate, though annoying.

The defecation bags containing a germicide to prevent bacteria and gas formation were easily sealed and stored in empty food containers in the equipment bay. It was filling them that was difficult. That took crew members forty-five minutes to an hour, during which time all work onboard ceased. Consequently the crew had a total of only twelve defecations between them over a period of nearly eleven days.

Urination was much easier as it did not involve removing clothing. There was a collection service inside both the pressure suits and the in-flight coveralls. Both devices could be attached to the urine dump hose and emptied into space. They had half expected the hose valve to freeze up in the vacuum, but it never did.

The batteries needed for reentry after the SM with its fuel cells was jettisoned, returned between fifty and seventy-five percent less energy than expected. More serious was the overheating of fuel cells, which might have failed when the spacecraft was too far from Earth to return on batteries, even if fully charged. These problems would have to be solved before Apollo 8 flew.

## CATCHING A COLD IN SPACE

Another aim of the mission was to discover whether a three-man crew could work together in these conditions. During the mission physical discomfort made them grumpy. Then, about fifteen hours into the flight, Schirra developed a bad cold. Cunningham and Eisele soon followed suit. A cold is uncomfortable enough on the ground, but in weightlessness it presents a different problem.

As mucus accumulates, it fills the nasal passages and does not drain from the head. The only relief is to blow hard, which is painful to the ear drums. So the crew of Apollo 7 took aspirin and decongestant tablets, and discussed their symptoms with doctors.

## Putting Apollo on TV

One piece of equipment got aboard in spite of the insistence of most engineers that it was not required. This was the television camera. Ever since September 1963, when NASA had first directed North American to install a portable camera in the spacecraft, the device had been going in and out of the craft as engineers wrestled with the constant problem of overweight. On occasion when kilograms, and even grams, needed to be saved from the command module, the camera was the first item to go. There were those in NASA, however, who persistently argued for the inclusion of television. In the spring of 1964, William A. Lee, Director of Systems Studies at the Houston Apollo office wrote:

> *One [objective] of the Apollo Program is to impress the world with our space supremacy. It may be assumed that the first attempt to land on the Moon will have generated a high degree of interest around the world ... A large portion of the civilized world will be at their TV sets wondering whether the attempt will succeed or fail. The question is whether the public will receive the report of this climactic moment visually or by voice alone.*

Finally in 1968 following four years of indecision, it was agreed to install television on Apollo 7. The revelation of the fuzzy, blurry, black-and-white pictures that were sent back from space made such compelling viewing that TV immediately established itself as an essential part of all subsequent manned space missions.

Astronauts Walter Schirra (right) and Donn Eisele seen in the first live television transmission from space.

Several days before the mission ended, they began to worry about wearing their spacesuit helmets during reentry, which would prevent them from blowing their noses. The build-up of pressure might burst their eardrums. Deke Slayton in Mission Control tried to persuade them to wear the helmets anyway, but Schirra was adamant. They each took a decongestant pill about an hour before reentry and made it through the acceleration zone without any problems with their ears. It was later discovered that an influenza epidemic at the Kennedy Space Center at the time had been responsible for infecting the crew.

## THE MAGNIFICENT FLYING MACHINE

The SPS engine, which had to fire the CSM into and out of the Moon's orbit, worked perfectly during eight burns lasting from half-a-second to 67.6 seconds. Apollo's flotation bags had their first tryout when the spacecraft splashed down in the Atlantic south-east of Bermuda, around a mile (1.6 km) from the planned impact point. The module turned upside down, but when inflated, the brightly colored bags flipped it upright. The crew and then the capsule were picked up by helicopters and deposited on the deck of USS *Essex*.

Apollo 7 had accomplished what it set out to do. It had checked out the CSM, giving the green light for the lunar orbit mission to follow. Its success also attracted media interest again. The crew underwent a six-day debrief for the benefit of the crew of Apollo 8, whose mission objectives were reviewed in the light of the outstanding accomplishment of the magnificent flying machine—Apollo 7. Partial to a good pun, the media described the achievement of the mission as "out-of-this-world!"

## Soviet Lunar Flyby

Despite the Apollo 7 triumph, the Russians seemed to be edging ahead in the race to the Moon. On September 16, 1968, the Soviet's Zond 5 circumnavigated the Moon carrying two tortoises, mealworms, wine flies, plants, seeds, and bacteria. Everything onboard was killed during reentry, but the Soviets had achieved the first lunar flyby and the problems could be fixed. Wernher von Braun was convinced that the Soviets were likely to win the race to the Moon as they were spending more money.

The following month Soyuz 3 carrying 47-year-old World War II veteran Georgy Beregovoy rendezvoused with the unmanned Soyuz 2 and landed safely back on the steppes of Kazakhstan, though it had failed to dock—a vital maneuver for any manned mission to the Moon. Then in December 1968, Zond 6 orbited the Moon, but crashed on its return to Earth when the parachute failed. James Webb too believed that the Soviets were winning the space race, though NASA was accused of exaggerating the Soviets' achievements to protect their budgets.

The Zond 5 automatic interplanetary station was the first spacecraft to orbit the Moon and return to Earth.

# CHAPTER 13

# THE MOST FANTASTIC VOYAGE OF ALL TIME

## APOLLO 8

Frank Borman volunteered to command Apollo 8 on a manned circumnavigation of the Moon, though he told Chris Kraft that he believed that they only had a fifty-fifty chance of making it home. Neil Armstrong and the other astronauts thought this was a bold move, but it would put America back ahead in the space race.

With only weeks to go before the scheduled launch of Apollo 8, problems were found with the J-2 engines that powered stage two and three. These were fixed with the brainstorming of 400 technicians and 31,000 man-hours of investigation before Apollo 8 was passed good to go.

Apollo 8 was supposed to carry LM-3 for tests in Earth orbit, but the problem of the LM's windows shattering under pressurization had still to be solved. To understand the failure, every shard was collected from inside the vacuum chamber and the pane of glass was painstakingly pieced back together. It was found that the glass had absorbed moisture during manufacture. This expanded as the pressure dropped in the vacuum chamber, causing the window to shatter. More stringent tests were introduced to detect defective glass.

However, inspection of the LM-3 found over 200 other defects. It would not be ready for the first manned flight, so Apollo 8 was redesignated as a CSM-only flight and Grumman would have another three months to fix the LM, which would then fly on Apollo 9.

### THE FASTEST HUMANS EVER

Apollo 8 lifted off on December 21, 1968. On the first night in space, Borman got ill and vomited—with particularly unpleasant consequences in zero-g. Crew members James Lovell and William Anders had to go around with paper wipes, trying to collect particles of flying puke. Fortunately, whatever was affecting Borman did not prove contagious. The nausea seems to have been related to sleeping pills he had been taking.

Then after 2 hours 38 minutes of orbiting the Earth, all tests were completed and Apollo 8 fired up its translunar injection burn that sent it on its way toward the Moon. The third stage of the Saturn V burned for just 318 seconds, accelerating it to 24,208 mph (38,959 kph)—the highest speed humans had ever traveled relative to the Earth. This was just less than Earth's escape velocity to put Apollo 8 into an elongated elliptical orbit, taking it to the point where the gravity of the Moon would capture it.

On their way, the crew were the first humans to travel through the Van Allen radiation belts that extend 15,000 miles (24,000 km) from Earth discovered by Explorer 1 and Explorer 3, but it was estimated that the astronauts would pass through them so quickly that they would suffer no more radiation than having a chest X-ray.

## LONGEST FOUR MINUTES OF THEIR LIVES

The spacecraft went into an orbit around 70 miles (111 km) above the surface of the Moon. It made ten orbits over twenty hours. At its furthest, it was 234,474 miles (377,349 km) from the Earth. Communication was lost as the CM went behind the Moon where the engine had to be fired for 4 minutes 7 seconds to slow the spacecraft into lunar orbit.

The crew described this as the longest four minutes of their lives. If the burn went on too long, the spacecraft would crash into the Moon; too short a burn would put them in a highly elliptical orbit, or would even fling them out into space. On Earth, Mission Control also had its nail-biting moments when Apollo 8 did not appear from behind the Moon at the exact moment predicted.

When eventually the spacecraft did emerge, Anders took two shots of the marbled blue Earth rising over the lunar landscape which would become world famous. Apollo 8 was scheduled to be in orbit around the Moon on Christmas Eve, so NASA asked the crew to prepare a special Christmas message. On December 24, the TV camera onboard broadcast the image of the Earthrise back to Earth, while Borman described what he could see below. Then Anders, Lovell, and Borman took turns reading the first verses of the Book of Genesis:

> *In the beginning, God created the heaven and earth. And the earth was without form, and void; and darkness was upon the face of the deep. And the Spirit of God moved upon the face of the waters. And God said, Let there be light: and there was light …*

Borman concluded by wishing everyone on Earth a Merry Christmas from the crew of Apollo 8, saying:

> *And from the crew of Apollo 8, we close with good night, good luck, a Merry Christmas and God bless all of you—all of you on the good Earth.*

It was estimated that a billion people were watching on television or listening on the radio. At the time, it was the most-watched TV broadcast ever. There were six telecasts during the mission, shown on all five continents.

## CHRISTMAS ON APOLLO

After ten orbits, on the morning of Christmas Day, Borman reignited the Apollo engine to boost the spacecraft out of the gravitational pull of the Moon. As Apollo sped back toward the Earth, Lovell maneuvered the spacecraft to take navigational sightings from various stars. In the process, he accidentally erased some data from the onboard computer guiding the Inertial Measurement Unit, or IMU, that corrected the craft's attitude. When he realized what had happened, it took fifteen minutes to reenter the information.

That afternoon, after making a transmission showing the audience on Earth the rudiments of life in space, they tucked into the same individual turkey dinners issued to front-line troops in Vietnam. However, their Christmas lunch was supplemented with brandy. Borman recalled:

> *When we opened up the dinner for Christmas and I found somebody had included brandy in there, you know, I didn't find it funny at all … If we'd have drunk one drop of that brandy and the thing would have blown up on the way home, they'd have blamed the brandy on it.*

As they approached the Earth, the CM was separated from the SM. They were then committed to reentry. To

One of the famous Earthrise pictures taken by Bill Anders during the Apollo 8 mission, showing the Earth rising above the lunar horizon. The land mass visible just above the terminator line is West Africa. This phenomenon is only visible to an observer in motion relative to the Moon. The two craters visible on the image were named 8 Homeward and Anders' Earthrise in honor of the Apollo 8 mission.

slow the spacecraft during reentry, it altered attitude so it skipped across the edge of the atmosphere. This subjected the crew to a deceleration of up to 6g. The reentry velocity was 24,696 mph (39,944 kph), with the heat shield reaching a temperature of 5,000°F (2,760°C).

The parachute deployed and Apollo 8 splashed down 147 hours after launch, precisely on time, only 2.5 miles (4 km) from the recovery ship USS *Yorktown*.

## THE BIG BIG BANG

After splashdown, Frank Borman expressed his shock and surprise at the impact when they came down to Earth, saying:

> *We hit the water with a real bang! I mean it was a big, big bang! And when we hit, we all got inundated with water. I don't know whether it came in one of the vents or whether it was just moisture that had collected on the environmental control system … Here were the three of us, having just come back from the Moon, we're floating upside down in very rough seas—to me, rough seas. Of course, to the consternation of Bill and Jim, I got good and seasick and threw up all over everything at that point.*

William Anders sounded like he had just about had enough of living at such close quarters with his two astronaut colleagues, when he recalled :

> *By now the spacecraft was a real mess, you know, not just from him but from all of us. You can't imagine living in something that close; it's like being in an outhouse and after a while you just don't care …*

This Apollo 8 reentry photograph was taken by a US Air Force Airborne Lightweight Optical Tracking System (ALOTS) camera mounted on a KC-135-A aircraft flown at 40,000 feet altitude.

There was a 10-foot (305-cm) swell and it took some time for the floatation balloons to right the spacecraft, and almost forty-five minutes for divers from USS *Yorktown* to arrive. After another forty-five minutes the capsule was on the deck of the aircraft carrier.

*The New York Times* called Apollo 8 "the most fantastic voyage of all time" and *Time* magazine made the crew its "men of the year." For a moment, America seemed to have edged ahead in the space race. But not for long.

On January 16, 1969, Soyuz 4 and Soyuz 5 became the first two spacecraft to dock in orbit. Then two cosmonauts transferred from Soyuz 5 to Soyuz 4. As no transfer tunnel had been developed, this had to be done by a spacewalk. The two spacecraft then undocked and made a safe return to the Earth. The two competitors in the space race were now neck and neck.

> *Apollo 8 has 5,600,000 parts and one-and-a-half million systems, subsystems, and assemblies. Even if all functioned with 99.9 percent reliability, we could still expect 5,600 defects.*
>
> NASA Safety Chief Jerry Lederer

## APOLLO 9

By March 3, 1969, the US was back in contention in the space race, with Apollo 9 carrying the LM into space. It would be the first time the entire Apollo spacecraft would fly. The crew were James A. McDivitt, David R. Scott, and Russell L. Schweickart, and the mission was to last ten days. During that time, they put LM-3 through its paces. Vitally, they had to practice docking and undocking of the CSM and LM.

Initially, the spacecraft went into an Earth orbit 117 by 119 miles (188 by 191 km). On the first day, the venting of the S-IVB propellant tanks changed the orbit to 123 by 127 miles (200 by 204 km). The CSM separated and the translation thrusters carried it a safe distance away, and the panel walls of the SLA that housed the LM-3 were jettisoned.

## DOCKING IN SPACE

The CSM's rotation thrusters then flipped it 180° and thrusters pushed it back toward the LM on the top of the S-IVB. A probe on the top of the CSM was aimed into a hole on the top of the LM. Once it was located, three small capture latches closed into the locked position. This in turn activated a probe retract system that drew the two vehicles together. At the same time, a tunnel ring on the LM activated twelve latches on the CSM probe to form a tight pressure seal between the two spacecraft.

The CSM-LM craft then separated from the S-IVB and the thrusters placed the CSM-LM a safe distance from the S-IVB which was then put through its paces. A sixty-two second restart of its engine raised the apogee to 1,895 miles (3,050 km). To achieve the planned escape trajectory, the S-IVB restarted a second time for 4 minutes 2 seconds. This resulted in less than the desired velocity. It did not affect Apollo 9, a lunar mission which might have had to be aborted.

Before the third S-IVB burn, the CSM's SPS engine fired for five seconds, placing the CSM in an orbit of 125 by 145 miles (201 by 233 km). This checked the capability of the guidance and navigation system to control the burn and tested the LM's ability to withstand acceleration and vibration. Three more burns of the SPS engine tested the structural dynamics of the docked CSM-LM under the conditions of a simulated lunar mission.

## TRANSFERRING TO THE LM

On the third day of the flight, McDivitt and Schweickart put on spacesuits. A valve was then opened, allowing oxygen to pass from the CM into the LM. The hatch was opened and electrical connections were made. McDivitt and Schweickart then transferred to the LM through the tunnel connection to perform a systems checkout. This included a 367-second firing of the LM descent engine to simulate the throttle pattern to be used during a lunar landing mission. McDivitt controlled the final fifty-nine seconds, varying the thrust from ten to forty percent before shutting it off manually. This was the first crewed throttling of an

Top: Apollo 9 CSM *Gumdrop* photographed through the window of the LM *Spider*. Bottom: The LM *Spider* in a lunar landing configuration as photographed from CSM *Gumdrop*.

Astronaut David R. Scott's stand-up extravehicular activity on the fourth day of the Apollo 9 mission. Scott is standing in the open hatch of the docked Command Module with the Earth in the background.

engine in space and increased the spacecraft's orbit to 130 by 300 miles (209 by 483 km).

After about nine hours, McDivitt and Schweickart transferred back to the CSM. The following day McDivitt and Schweickart reentered the LM. As Schweickart was suffering from nausea, his scheduled two-hour EVA to simulate external transfer rescue techniques was curtailed. Instead, he climbed out of the LM for a 37.5 minute EVA, testing the EVA suit. Unlike early models which depended for life support on an umbilical cord attached to the mother ship, this one had an independent life-support system built into a backpack.

## THE LM LEARNS TO FLY

The next day, McDivitt and Schweickart boarded the LM again. This time they separated from the CSM. This was the first manned spaceflight of a craft not equipped to reenter the Earth's atmosphere.

They fired the LM's descent engine once for 24.9 seconds to place the spacecraft into an orbit 137 by 167 miles (220 by 268 km). It fired again for 24.4 seconds to circularize the orbit to about 154 by 160 miles (247 by 257 km), some 12 miles (19 km) higher than the CSM. Four hours later, with the LM 113 miles (182 km) away from the CSM, it jettisoned the descent stage and fired the ascent stage engine for the first time in space. This lowered the LM orbit by 11 miles (17 km) and placed it 75 miles (120 km) behind and 10 miles (16 km) below the CSM. From there, it began a rendezvous. Six hours later, the CSM and LM redocked. Once the crew had transferred back into the CSM, the LM ascent stage jettisoned and was commanded remotely to fire its engine until it ran out of fuel.

## COMING HOME

Two telecasts were made to Earth from Apollo 9. The first from inside the CM lasted almost seven minutes. The second telecast lasted about thirteen minutes and showed interior views of the LM. Photographs were also taken as part of the multi-spectral terrain photographic experiment.

On the sixth, seventh, and eighth day, the SPS engine was restarted three more times. On the tenth day, March 15, the SPS engine fired one last time to take the CSM out of orbit, after a delay of one orbit because of high seas in the recovery area. The SM was then jettisoned and the CM splashed down some 341 miles (548 km) north of Puerto Rico, 3 miles (5 km) from the recovery ship.

The flight totaled 241 hours 53 seconds—ten seconds longer than planned. The S-IVB stage remained in orbit around the Sun, while the LM ascent stage reached Earth orbit. The LM descent stage decayed on March 22, 1969. The ascent stage's orbit eventually decayed on October 23, 1981.

## APOLLO 10

Launched on May 18, 1969, Apollo 10 was to be a full-scale dress rehearsal for the Moon landing. The crew were Thomas P. Stafford, John W. Young, and Eugene A. Cernan.

The Saturn V took Apollo 10 into a 115-mile (185-km) diameter circular parking orbit around the Earth. One-and-a-half orbits later, the S-IVB fired to increase velocity from 17,459 mph (28,098 kph) to 24,989 mph (40,216 kph) on a free-return trajectory. This is a trajectory that, without any further thrust from the engines or alteration in course, would have the spacecraft loop around the Moon under its gravitational pull and return it to Earth.

After twenty-five minutes, the CSM separated for transposition and docking with the LM, similar to the maneuver performed on Apollo 9.

The Apollo 10 Command and Service Module, *Charlie Brown*, is photographed from the Lunar Module, *Snoopy*, after separation in lunar orbit. Numerous bright craters and the absence of shadows show that the sun was almost directly overhead when this photograph was taken.

Approaching from below, the ascent stage of LM Snoopy is photographed from CSM *Charlie Brown* prior to docking in lunar orbit.

## GOING LIVE ON TV

The first live color TV transmissions to Earth began three hours after launch when Apollo 10 was 3,570 miles (6,035 km) from Earth and showed the docking process and the interior of the CM. It concluded when the spacecraft was 9,428 miles (15,172 km) away.

About four hours after the launch, Apollo 10 separated from the S-IVB, which went into orbit around the Sun. The separation was followed by another telecast from 14,625 miles (23,536 km) out. A third TV transmission showing pictures of the Earth was made from 24,183 miles (38,918 km) out, and a fourth telecast again showing the Earth was made from 140,000 miles (225,308 km).

The launch trajectory had been so satisfactory that only one of a planned four midcourse corrections was needed. This was done 26 hours 30 minutes into the flight. It took three days to reach the Moon.

## FLY ME TO THE MOON

About seventy-six hours into the flight, the SPS engine was fired putting the craft into lunar orbit. A second firing of the SPS 4 hours 30 minutes later circularized the lunar orbit of Apollo 10 at an altitude of approximately 69 miles (111 km). Then the first color TV pictures of the Moon's surface were transmitted back to Earth.

On the twelfth orbit of the Moon, Stafford and Cernan entered the LM and prepared for the undocking maneuver, while Young remained in the CM. The vehicles separated and briefly flew a station-keeping lunar orbit of 66.7 by 71.5 miles (107 by 115 km).

To achieve a simulation of the future Apollo 11 landing, the LM descent engine fired for 27.4 seconds, with ten percent thrust for the first fifteen seconds and forty percent thrust for the rest. This brought the LM to a new orbit of 9.7 by 70.5 miles (15.6 by 113.5 km). This was the altitude the powered descent would start on Apollo 11. The LM only carried enough fuel to lift it back into orbit from that height, not from the surface of the Moon so the astronauts would not be tempted to make an unplanned landing.

The LM flew over Apollo 11's intended landing site in the Sea of Tranquility. During this run, the LM landing radar was tested. Following a 7.5-second firing of the LM reaction control system (RCS) thrusters, the descent engine fired in two bursts for 40.1 seconds—at ten percent and at full throttle.

On the fourteenth revolution, the ascent stage separated from the descent stage, but when it was jettisoned on a second attempt, the ascent stage began to roll violently. This was later attributed to an error in a flight-plan checklist, causing an incorrect switch position. The pilot had inadvertently duplicated the commands of the flight computer. The problem was soon rectified. The descent stage remained in orbit around the Moon.

The ascent engine fired for fifteen seconds, lowering the LM maximum altitude of the orbit to 53.8 miles (86.6 km), putting the LM below the CSM and 230 miles (370 km) behind it. The RCS thrusters fired for 27.3 seconds when the LM was 16.9 miles (27.2 km) below the CSM and 170.4 miles (274.2 km) behind it. Preparing for rendezvous, the RCS fired again.

## RENDEZVOUS IN SPACE

Stafford sighted the CSM's running lights at a distance of about 48 miles (75 km). The fifteen-second firing reduced the velocity as the LM entered an intercept trajectory and the two vehicles achieved station-keeping at the sixteenth orbit of the Moon. With Young in the CSM taking on an active rendezvous role, the vehicles were redocked slightly more than 106 hours into the mission.

Stafford and Cernan then transferred back into the CM, sealing the hatch behind them. The LM ascent stage was then jettisoned. Its engines were fired remotely and it headed into orbit around the Sun. In later missions, the LM was crashed into the Moon to get readings from the seismometers left on the surface.

The CSM continued in lunar orbit, tracking landmarks and taking photographs. On the thirty-first orbit, the SPS restarted on the far side of the Moon, putting it into trans-Earth trajectory.

After a midcourse correction, the CSM separated, Apollo 10 reentered Earth's atmosphere on May 26. The module splashed down in the Pacific Ocean within television range of its primary recovery ship, USS *Princeton*. Apollo 10 completed a flight of 192 hours 3 minutes 23 seconds—1 minute 24 seconds longer than planned.

PART FIVE

# ONE SMALL STEP

*THAT'S ONE SMALL STEP FOR (A) MAN.*
*ONE GIANT LEAP FOR MANKIND.*

MISSION COMMANDER NEIL ARMSTRONG
THE FIRST MAN ON THE MOON

CHAPTER 14

# FIRST MEN ON THE MOON

## APOLLO 11

At last, in July 1969, NASA was ready to fulfill President Kennedy's promise. Apollo 11 launched from Cape Kennedy on July 16, carrying Commander Neil Armstrong, Command Module Pilot Michael Collins, and Lunar Module Pilot Edwin "Buzz" Aldrin. All three had been into space before.

NASA estimated that one million people turned up at Cape Kennedy to watch the launch. They brought campers, pitched tents on the beach, or slept in their cars. More than 3,400 reporters from newspapers and TV stations assembled at Cape Kennedy and the Houston control center. Among them were 700 correspondents from overseas, with the largest contingent of 111 coming from Japan. Among the dignitaries on hand were former President Lyndon Johnson and Charles Lindbergh, the first man to make a solo flight across the Atlantic.

The crew were awoken at 4:00 a.m., ready to board Apollo 11 at 7:00 a.m. After a medical check, they were given the traditional steak and eggs breakfasts before donning their bulky metallic space suits.

At around 6:30 a.m. they boarded the van that would take them the 5 miles (8 km) to the launch site, Launch Pad 39A. An hour before liftoff the spacecraft was closed off and the remaining technicians cleared from the site. The access arm from the tower was retracted.

### LIFTOFF

With less than thirty minutes to go the spacecraft switched to internal power, and final communications tests were performed. In the final six minutes of the countdown, tracking stations around the world were alerted that the launch was about to take place.

The final firing command was given 3 minutes 10 seconds before the launch. From then on, the automatic sequencing system took over and no one at the control center could halt the launch.

At 9:32 a.m. the swing arms move away and a plume of flame signals the liftoff of the Apollo 11 Saturn V space vehicle and astronauts Neil Armstrong, Michael Collins, and Buzz Aldrin from Kennedy Space Center Launch Complex 39A.

With 8.9 seconds to go, the first-stage engines were ignited. Two seconds before liftoff, with all the first-stage engines running, the thrust built up to 7.6 million pounds and the huge rocket began to lift slowly off the launch pad.

The astronauts found themselves pushed down in their seats. They were pulling 3g as the rocket disappeared from sight, leaving a trail of smoke and a distant roar. At 43,000 feet (13,000 m), they reached 1,800 mph (2,900 kph). It is there that they hit maximum dynamic pressure. The velocity combined with the atmospheric density put the structure under intense stress and the rocket now pulled 4.5g. The astronauts themselves felt an intense vibration, which made even flicking a switch a difficult task. At 200,000 feet—37 miles or 61 km—above the Earth they reached 5,300 mph (8,600 kph).

## A GIANT ROARING FIREBALL

Just 2 minutes 40 seconds into the flight, the Saturn V's 138-foot (42 m) first stage was exhausted, having burned 4.5 million pounds—over 2,250 tons—of fuel. At 217,000 feet (66,000 m), the speed had reached 6,100 mph (9,800 kph) when the first stage's five engines shut down. Eight retro-rockets fired to separate the first stage from the rest of the rocket. It fell away to splash down in the Atlantic.

The immense thrust of the first stage had compressed the ship above it. The ship then decompressed, taking the crew from 4g to –1.5g in just over a second. At the same time, a giant fireball came roaring up the length of the booster passing the capsule. The second stage then fired pushing the capsule through the fireball.

One second after separation, four ullage motors with 21,000-pounds of thrust each fired for five seconds to settle the zero-g fuel in the tanks. The five J-2 engines then fired for six minutes, taking Apollo 11 to 600,000 feet (183,000 m) and a speed of over 10,000 miles an hour (16,500 kph).

The LET was jettisoned 3 minutes 17 seconds after liftoff. It could do no good now. Then the crew had vision out of the capsule windows, but the only thing they could see was the blackness of the sky.

Another 10,000 feet (3,000 m) and the speed rose to 15,500 mph (25,000 kph). The second stage's rockets then fell silent. The retro-rockets fired and the spent second stage fell away. Then the third stage's single J-2 engine fired. This took them into orbit at an altitude of 116 miles (187 km).

Members of the Kennedy Space Center team rise from their consoles to watch the Apollo 11 liftoff.

## TRANSLUNAR INJECTION

After 2 hours 44 minutes, Mission Control in Houston gave the go-ahead for translunar injection. Stage three's J-2 engine reignited and fired for 5 minutes 47 seconds, boosting the speed to 24,182 mph (38,917 kph).

At 500 miles (805 km) from Earth, Command Module Pilot Michael Collins fired the retro-rockets that detached the capsule from stage three. It sped away at a relative velocity of half-a-mile-an-hour, though both were traveling at 17,000 mph (27,360 kph) relative to the Earth.

The leaves of the SLA then opened, exposing the LM. Collins used a rangefinder eyepiece with crosshairs to locate the probe on the CSM into the drogue on the LM. Once the computer had confirmed that the alignment was correct, he fired the latches that locked the two craft together. The LM was then freed from the third stage of the Saturn V, which fired once more to slingshot it past the Moon into solar orbit.

Once the CSM's probe and the LM's drogue was removed, it was possible to equalize the pressure between the two craft. Aldrin then squeezed through the hatch into the LM, designated *Eagle* on this mission, to inspect it for any damage that may have been caused on launch. He found the sensation intriguing, especially the weightlessness:

> *As you climbed, seemingly upward through the ceiling of the Command Module, Columbia, you emerged through the ceiling of the LM—upside-down, or so it seemed.*

## PREPARING FOR LANDING

Collins took over the running of the CSM—charging batteries, renewing carbon-dioxide filters, expelling waste, and checking drinking water—while Armstrong and Aldrin prepared the LM for landing. Then they tried to get some sleep in mesh hammocks that stopped them knocking into dashboard switches, or each other. They woke to coffee, sausages, cinnamon toast, and fruit cocktail.

During the translunar injection, Collins spent his time squinting through the ship's sextant to make navigational corrections. These were done by adjusting the Service Module's rocket on its gimbal. The same engine was then used to slow the craft so that it would fall into lunar orbit under the Moon's gravity.

As the gravity of the Moon took over, the Earth's gravity had slowed the craft to 3,400 mph (5,500 kph). After that, it speeded up again. Collins then turned the capsule so they could see their destination—the Moon.

At nearly a day-and-a-half from liftoff, they gave their first global TV broadcast. Collins showed the world their food packages, while Armstrong pointed out the geography of the Earth that could be seen below.

For Apollo 11, it was necessary to achieve a near perfect circular orbit around the Moon for *Eagle* to rendezvous with *Columbia* once the Moon landing was over. As the spacecraft approached the Moon, the CSM engine burned for 6 minutes 2 seconds, slowing it to 2,000 mph (3,220 kph). The gravitational pull of the Moon then hauled Apollo around the far side and radio communication blacked out.

When the CSM came around the far side of the Moon and radio communication was reestablished, it was clear that the burn of the SPS engine had been of exactly the right duration to put the craft into lunar orbit. Otherwise the mission would have been aborted.

The craft then rolled over and pitched down so the crew could see the surface. They had not been able to see the Moon for nearly a day. Collins said:

> *All of us are aware that the honeymoon is over and we are about to lay our little pink bodies on the line.*

## THE *EAGLE* HAS WINGS

After thirteen orbits of the Moon, Armstrong and Aldrin put on their EVA suits. Collins also wore a lighter suit in case there was a problem with the undocking procedure.

Armstrong and Aldrin took up their positions in the LM. Inside there were no seats. The LM had to be flown standing up with a dashboard stick to control yaw, pitch, and roll. The crewmen's boots were Velcroed to the carpet and their waists tethered to the walls.

Once the hatch was sealed, Collins severed the electrical connections and reassembled the docking probe and drogue. However, he decided not to take time setting up a TV camera to film the undocking. The three astronauts then ran through their checklists, preparing for separation.

At 1:47 p.m. CST on July 20, on the far side of the Moon, Collins fired the explosives that separated the

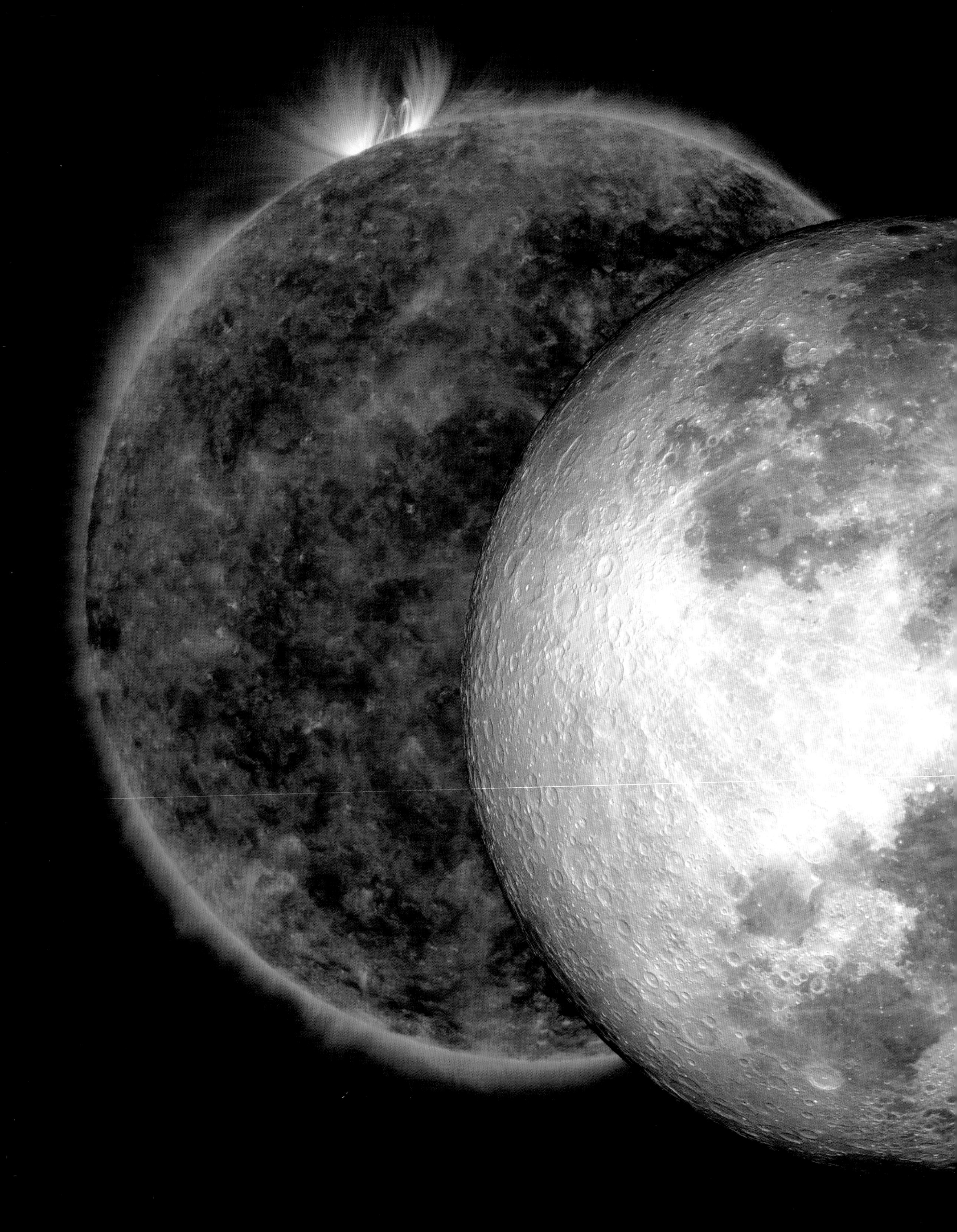

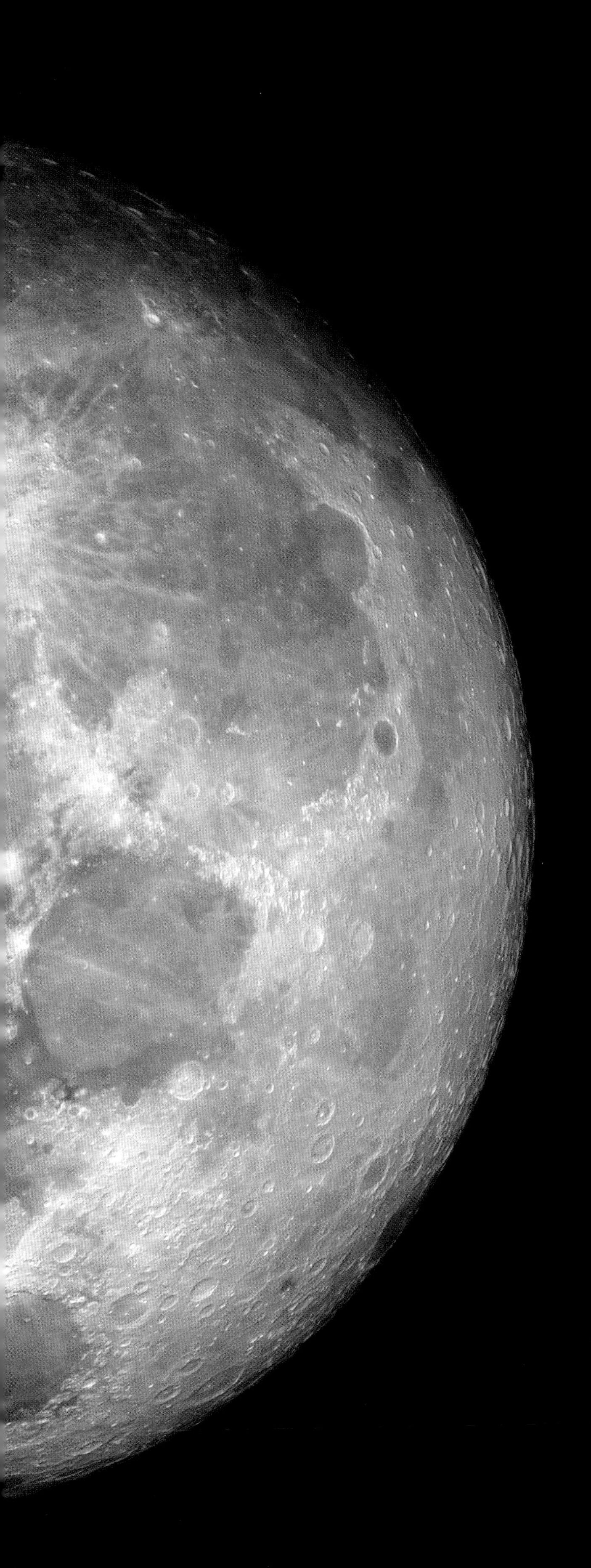

# APPROACHING THE MOON

Recalling the approach, Neil Armstrong said:

*The most dramatic recollections I had were the sights themselves. Of all the spectacular views we had, the most impressive to me was on the way to the Moon, when we flew through its shadow. We were still thousands of miles away, but close enough so that the Moon almost filled our circular window. It was eclipsing the Sun, from our position, and the corona of the Sun was visible around the limb of the Moon as a gigantic lens-shaped or saucer-shaped light, stretching out to several lunar diameters. It was magnificent, but the Moon was even more so. We were in its shadow, so there was no part of it illuminated by the Sun. It was illuminated only by earthshine. It made the Moon appear blue-gray, and the entire scene looked decidedly three-dimensional. I was really aware, visually aware, that the Moon was in fact a sphere, not a disc. It seemed almost as if it were showing us its roundness, its similarity in shape to our Earth, in a sort of welcome. I was sure that it would be a hospitable host. It had been awaiting its first visitors for a long time.*

The roundness of the Moon, the "hospitable host" that Neil Armstrong saw from Apollo 11, with its crisp horizon can be clearly seen as it eclipses the Sun.

*Eagle* from *Columbia*. The LM then swiveled around so that Collins could check that the telescopic legs had deployed correctly.

Finally the two spacecraft came around the Moon and Mission Control in Houston heard Armstrong say: "The *Eagle* has wings."

Collins then said to Armstrong: "You've got a fine-looking flying machine there, *Eagle*, despite the fact you're upside down."

Something had gone wrong though. Before separation the tunnel between the CSM and the LM was supposed to have been depressurized until it was a vacuum. Instead a little air had been left in it and the pressure gave *Eagle* a little extra speed, throwing out NASA's finely calculated landing plan. This meant they could not depend on the onboard computer to control the landing. Armstrong would have to do it himself.

## SEA OF TRANQUILITY

NASA had selected an area in the south-west corner of the Sea of Tranquility for the first lunar landing. Flyby photos had shown that it was relatively free of craters and boulders. It lay beneath *Columbia*'s orbit and the Sun would be behind them at a low ten-degree angle so they would not be hindered by too much glare or shadow. For just one day each lunar month, this was the ideal landing place.

At an altitude of 10 miles (16 km), the *Eagle*'s rocket fired, slowing the LM to begin the powered descent. The onboard Raytheon computer used the landing radar to make the descent to around 500 feet (152 m). The computer then gimbaled the *Eagle*'s rocket and fired its thrusters, approaching the Sea of Tranquility with the portholes facing downward so Armstrong could assess the terrain.

If he found the preprogrammed landing site unsuitable, he could use the Abort Guidance System (AGS) landing radar to land the *Eagle* himself. Armstrong had extensive training on the Lunar Landing Research Vehicle in Houston which simulated the Moon landing.

The *Eagle* was approaching the landing zone when Houston lost contact. Communications then had to be patched through *Columbia*, adding an awkward delay. This was crucial and the time was approaching when the decision had to be made to set down the craft or not. The experts in the Mission Operations Control Room consulted and decided the landing should go ahead. The message was conveyed to the *Eagle* via Michael Collins on *Columbia*. Aldrin replied: "We read you ... Roger."

## LOOKING FOR LANDMARKS

While Armstrong took the controls and searched for a landing site, Aldrin kept his eyes on the dashboard and called out readings from the displays and gauges. The *Eagle* was following the flight path US1, over the Maskelyne A crater, Boot Hill which was shaped like Italy, the Rima Maskelyne 1 canyon, Last Ridge hill, and the planned landing site in the Moltke crater. It quickly became clear that they were going to overshoot by around 4 miles (6.4 km). This was because they were flying at almost 14 mph (22.5 km) faster than expected. If that figure had had been much higher, the landing would have been aborted. Again it was decided to continue.

They were descending at 90 mph (145 kph) when Armstrong switched on the AGS, which gave their altitude as 33,500 feet (10,000 m). This disagreed with the computer's landing radar by 2,900 feet (884 m). Then an alarm went off and the computer flashed up an error code. The computer was overloaded.

While Mission Operations Control Room tried to figure out what to do, the computer kicked in again. At 25,000 feet (7,620 m), it tilted the *Eagle* upright with its feet pointing downward, ready for landing. Light reflected from the Earth shone through the porthole and the engine powered up from ten percent to full power to smooth the descent.

## SKIMMING THE SURFACE

At 4,000 feet (1,220 m), the *Eagle* slowed to 70 mph (112 kph). The experts in Mission Operations Control Room who had consulted again and again, finally gave the go-ahead for landing. At 2,000 feet (610 m), the *Eagle* was traveling at 14 mph (22.5 kph). Armstrong was now lost. The *Eagle* was too far off course and too close to the ground for the photos taken by Apollo 10 to be of any help.

He switched the controls to semiautomatic. This left the computer controlling the engine's throttle and the descent velocity, while he used the hand controller to

# LANDING THE LUNAR MODULE

It took twelve minutes for the LM to slow down from the equivalent of Mach 5 to zero, to land on a precise point on the lunar surface. The landing began at the low point of the LM's elliptical orbit 250 miles (400 km) east of the landing site. The Commander then started the computer's braking program. This worked out how to burn the descent engine to hit the landing site. The pilot then had to approve this.

The reaction control system engines were fired briefly to force the propellant into the bottom of the tank. Then the main engine ignited—first at ten per cent, then cranking up to full power. The first major burn would last 7 minutes 30 seconds, slowing the LM from 3,750 mph (6,035 kph) to 410 mph (660 kph). Slowing it would drop it to an altitude of 10,000 feet (3,048 m) as it approached the landing site.

Meanwhile the Lunar Module Pilot monitored the PNGS and AGS guidance systems to make sure they agreed with the tracking data sent by Mission Control. The computer also checked the readings data from its inertial guidance platform. During this part of the descent, the crew were looking out into space with their backs to the surface to give the landing radar a clear view.

At 4.3 miles (6.9 km) out, the computer tilted the LM into a more upright position and handed control over to the Commander. His eyes were guided by the gradations of the Landing Point Designator etched into the window in front of him. It was up to him to find a spot smooth and hard enough to land on. Any movement of the control stick and the computer would recalculate the flight path.

Once the LM was down to 500 feet (152 m), the computer accepted no further changes in the landing site, though the Commander could still fine-tune the LM's descent. The reaction control system thrusters could still be used to slow the rate of descent, while the Lunar Module Pilot called out both the forward and vertical velocity. It was possible to slow the descent to just a few feet per second.

Landings always took place away from the Sun, so the Commander could get some visual clues from the LM's shadow as it chased across the landscape. When the landing probes, extending from under the footpads, made contact, a blue light lit on the control panel. The engine was cut and the LM fell the last three or four feet to the surface.

redesignate the landing point by looking through a grid on the window that acted like a gunsight.

At 300 feet (91 m), the *Eagle*'s descending speed slowed to just 2.5 mph (4 kph). Meanwhile it kept skimming forward across the surface, while Armstrong sought a suitable place to land. All Aldrin could do was call out the altimeter readings and their horizontal speed. Armstrong said it was clear the system wanted to land them in a boulder field.

## FINDING A PLACE TO PARK

Armstrong was very tempted to land, but his better judgment took over. The craft was pitched level, so it would maintain its horizontal velocity. He then took over control of the throttle and flew the LM manually the rest of the way. By then the fuel was getting low. They had to land soon or fire the ascent engine and abort.

*I was surprised by the size of these boulders. Some of them were as big as small motorcars.*

Neil Armstrong

Mission Control could not understand why Armstrong was taking so long to land. They kept watching his forward velocity. At around 30 mph (48 kph), it remained too fast to land. Armstrong was concentrating too hard to explain that he was trying to avoid a rock field. But he only had sixty seconds' worth of fuel left.

Descending at one foot per second, the LM began to slide sideways, then go backward. Armstrong recalled:

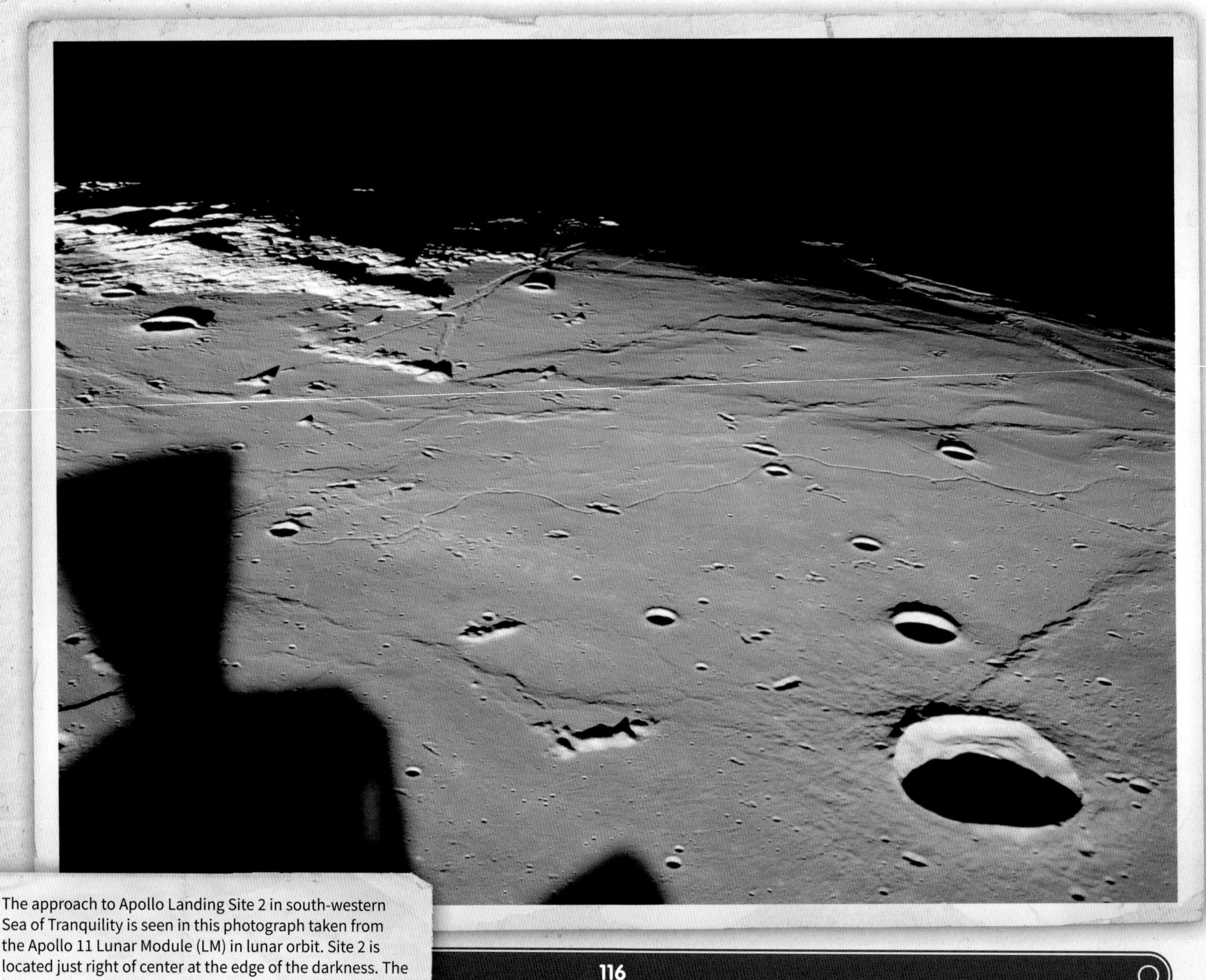

The approach to Apollo Landing Site 2 in south-western Sea of Tranquility is seen in this photograph taken from the Apollo 11 Lunar Module (LM) in lunar orbit. Site 2 is located just right of center at the edge of the darkness. The large crater at the lower right is the Rima Maskelyne 1.

*For some reason I'm not sure of, we started to pick up left translational velocity and a backward velocity. That's the thing I certainly didn't want to do, because you don't like to be going backward, unable to see where you're going. So I arrested this backward rate with some possibly spasmodic control motions, but I was unable to stop the left translational rate. As we approached the ground, I still had a left translational rate which made me reluctant to shut the engine off while I still had that rate. I was also reluctant to slow down my descent rate any more than it was, or stop [the descent], because we were close to running out of fuel. We were hitting our abort limit.*

## DEAD MAN'S ZONE

They were still around 100 feet (30 m) from the surface and Aldrin was getting concerned.

"We were in the so-called dead man's zone, and we couldn't remain there long," he said. "If we ran out of fuel at this altitude, we would crash into the surface before the ascent engine could lift us back toward orbit."

But he could hardly nudge Armstrong and say: "Neil, hurry up, get it on the ground."

Asked about it later, Armstrong said: "I was just absolutely adamant about my God-given right to be wishy-washy about where I was going to land."

While most of the people in Mission Control were concerned that Armstrong was going to run out of fuel, Engineering Analysis Office Chief Milton A. Silveria was more concerned that if the main engine fired close against a solid surface, the shock created would make the engine blow up.

The Mission Operations Control Room was designed for sixty people. There were about 150 in there for the landing, who were so silent you could have heard a pin drop. Goddard liaison Bill Easter said:

*That, to me, probably was the most exciting time I've ever had at NASA, when they started landing. There was no other feeling like that. It was like watching a man, some snake trainer, put his hand on a cobra, anything can happen any minute, and probably will.*

For Armstrong the search went on as if he was coolly driving round a parking lot on Earth:

*I changed my mind several times, looking for a parking place. Something would look good, and then as we got closer it really wasn't so good. Finally, we found an area ringed on one side by fairly good-sized craters and on the other side by a boulder field. It wasn't particularly big, a couple of hundred square feet—about the size of a big house lot.*

The landing was shown live on TV. Armstrong's wife Jan was watching, hugging their son Ricky to her, while Joan Aldrin supported herself by leaning against a door. Her hands were shaking and her eyes were wet with tears.

## ONLY THIRTY SECONDS OF FUEL

Onboard the *Eagle* Aldrin read off the altitude and descent speed: "Forty feet, down two and a half [feet per second] ... Picking up some dust ... Thirty feet, two and a half down."

They were down to thirty seconds of fuel.

"Four forward. Four forward. Drifting to the right a little," said Aldrin. "Twenty feet, down a half ... Drifting forward just a little bit; that's good."

Finally Armstrong saw a clear spot and righted the *Eagle* into the vertical position for landing. The flight surgeon noted his heart rate was 156; Aldrin's was 125. A normal adult's heart rate at rest is between sixty and a hundred.

The LM shuddered as the thrusters fired. At 30 feet (9 meters), the LM kicked up a dust cloud. Then, with about seventeen seconds of fuel remaining, a blue light on the dashboard lit up. One of the three-foot probes under the landing pads had made contact with the surface.

"Contact light," said Aldrin.

"Shutdown," said Armstrong and Aldrin cut the engine.

As the engine was shutdown, the LM fell the last three feet to the surface. Then Aldrin keyed "413" into the computer. This stored the craft's exact location—0.71° north, 23.63° east. Armstrong said later:

*I was absolutely dumbfounded when I shut the rocket engine off and the particles that were going out radially from the bottom of the engine fell all the way out over the horizon. They just raced out over the horizon and instantaneously disappeared, you know, just like it had been shut off for a week. That was remarkable. I'd never seen that. I'd never seen anything like that. And logic says, yes, that's the way it ought to be there, but I hadn't thought about it and I was surprised.*

He then reported to Mission Control: "Engine arm is off ... Houston, Tranquility Base here ... The *Eagle* has landed."

## LOST FOR WORDS

In Mission Control, people began cheering, clapping, and stomping their feet. But there was work to do. Flight Director Gene Kranz said:

> *We have to make sure, almost instantaneously, whether the spacecraft is safe to leave on the surface of the Moon or should we immediately lift off? We go through what we call our T-1 stay/no-stay decisions. So that within sixty seconds of getting on the Moon, I have to tell the crew, "It's safe to stay on the Moon for about the next eight minutes."*

But Kranz was so pumped up that he could not even speak. He was literally lost for words. He had to get going with the stay/no stay decision. In frustration at the cheering coming from the room, he rapped his arm down on the console, breaking a pencil. The CapCom (Capsule Communicator) Charlie Duke next to him looked at him in surprise. Finally, Kranz got back on track and, in a cracked voice, said: "Okay, all flight controllers, stand by for the T-1 stay/no stay."

> *Tranquility Base here ... The Eagle has landed.*
>
> Neil Armstrong

## PHEW! WOW! MAN ON THE MOON

Duke went back to work too. "Roger Twan ... Tranquility, we copy you on the ground," he said. He recalled later: "We were so excited I couldn't even pronounce Tranquility. It came out Twanquility."

"You've got a bunch of guys about to turn blue, we're breathing again, thanks a lot," he continued. "Be advised there are lots of smiling faces in this room, and all over the world."

"There are two of them up here," said Aldrin.

"And don't forget one in the command module," Collins added. "You guys did a fantastic job."

Walter Cronkite, commentating on the Moon landing, pulled off his glasses and mopped his brow. "Phew!" he said. "Wow, boy! Man on the Moon."

Armstrong and Aldrin shook hands and patted each other on the shoulder. But there was no time for reflection. No sooner had they landed, than they had to prepare for take-off in case of an emergency. Only later did Armstrong recall the elation he felt. That moment was, after all, what a third of a million people had worked for a decade to accomplish.

They began a countdown to test the ship's systems. If the mission was to be aborted, they would have to launch back to the mother ship right away, or risk waiting two hours for *Columbia* to complete another orbit.

## MISGUIDED AUTO-TARGETING

There was immediately a problem. On the cold surface of the Moon, propellant had frozen in one of the LM's pipes. The blockage led to a build-up of heat and pressure, threatening to cause an explosion. Engineers from Grumman reckoned that the engine could be switched on and off again quickly to "burp" the blockage out. But this in itself might topple the *Eagle* over. Meanwhile, Armstrong and Aldrin had to wait inside the ascent stage. Then, miraculously, the plug just melted away.

With the crisis over, Armstrong finally had a chance to report to Mission Control what had happened during the final descent:

> *Hey, Houston, that may have seemed like a very long final phase. The auto-targeting was taking us right into a football-field-sized crater, with a large number of big boulders and rocks for about one or two crater diameters around it, and it required us going in P66 and flying manually over the rock field to find a reasonably good area.*

P66 was the code for the rate of descent prior to landing. From his side of the craft, Aldrin could not see the danger, but Collins in *Columbia* could see just how rough the designated landing zone was and compared the surface to a corn cob.

"It really was rough, Mike," Armstrong said. "Over the targeted landing area, it was extremely rough, cratered, and large numbers of rocks that were probably ... larger than five or ten feet in size."

Aldrin was ready to give thanks. From his personal kit, he pulled out a chalice, a host, and few drops of red wine to take Holy Communion. This was not televised because the founder of the American Atheists, Madalyn Murray O'Hair, was already suing NASA over the Bible reading from Apollo 8.

## THOUSANDS OF LITTLE CRATERS

Both astronauts gave an account of what they saw. Aldrin told Mission Control:

*We'll get to the details of what's around here, but it looks like a collection of just about every variety of shape, angularity, granularity, about every variety of rock you could find. The color is … Well, it varies pretty much depending on how you're looking relative to the zero-phase point [the point directly opposite the Sun]. There doesn't appear to be too much of a general color at all. However, it looks as though some of the rocks and boulders, of which there are quite a few in the near area … It looks as though they're going to have some interesting colors to them.*

These were sights that no human had ever seen before and Neil Armstrong kept the descriptions to Mission Control coming:

*The sky is black. It's a very dark sky. But it still seemed more like daylight than darkness as we looked out the window. It's a peculiar thing, but the surface looked very warm and inviting. It looked as if it would be a nice place to take a sunbath. It was the sort of situation in which you felt like going out there in nothing but a swimming suit to get a little sun.*

*From the cockpit, the surface seemed to be tan. It's hard to account for that, because later when I held this material in my hand, it wasn't tan at all. It was black, gray and so on. It's some kind of lighting effect, but out the window the surface looks much more like light desert sand than black sand.*

The Apollo 11 landing area, as seen here from the LM Eagle, has been designated Statio Tranquillitatis which is the Latin equivalent of "Tranquility Base," and three small craters to the north of the base have been named Aldrin, Collins, and Armstrong in honor of the crew.

NASA
ARMSTRONG
NASA
COLLINS

The Apollo 11 lunar landing mission crew, pictured from left to right, Mission Commander Neil A. Armstrong, Command Module Pilot Michael Collins, and Lunar Module Pilot Edwin "Buzz" Aldrin Jr.

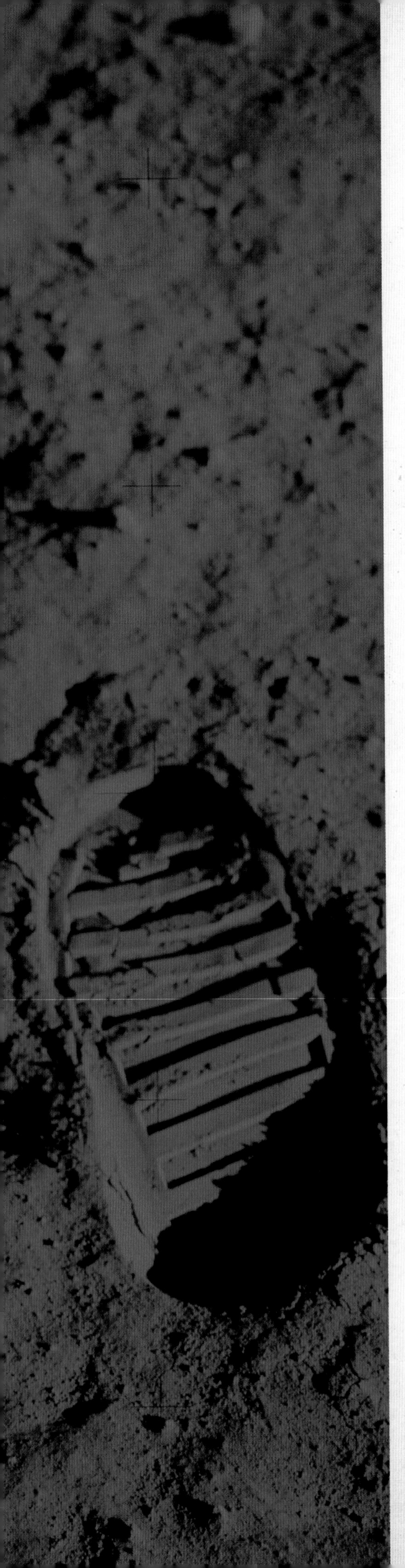



# CHAPTER 15

# WALKING ON THE MOON

The plan was for the astronauts to have a four-hour rest period before they went out to walk on the Moon. However, Armstrong and Aldrin did not feel like sleeping and requested that the rest period be postponed. Houston agreed, possibly because if the Moonwalk was made straight away, they would catch the primetime TV audience at 8:00 p.m. CST, rather than one in the morning. Armstrong said.

> *We had thought, even before launch, that if everything went perfectly and we were able to touch down precisely on time; if we didn't have any systems problems to concern us; and if we found that we could adapt to the one-sixth gravity lunar environment readily, then it would make more sense to go ahead and complete the EVA while we were still wide awake—but, in all candor, we didn't think this was a very high probability.*

Preparing for the Moonwalk was no easy task. The men had to disconnect themselves from the LM's life support systems and put on the backpacks containing their Personal Life Support Systems (PLSS) and their backup Oxygen Purge Systems. They then had to be connected to various outlets on their spacesuits. As the backpacks stuck out more than a foot, donning this bulky equipment within the confines of the tiny LM was difficult. "We felt like two fullbacks trying to change positions inside a Cub Scout pup tent," said Aldrin.

They filled their pockets with penlights, data cards, and scissors, then attached their boots, overshoes, and gloves to their spacesuits' metal cuffs. The whole operation was supposed to have taken two hours, instead it took three.

## MICROMETEORITES

All EVAs are dangerous. In space, astronauts are bombarded with solar radiation and cosmic rays. On the surface of the Moon there was an increased chance of being hit by a micrometeorite traveling at up to 64,000 mph (100,000 kph). The chances were estimated at 1 in 10,000. Consequently, the protection of a space suit was vital. The lunar EVA suit comprised three layers of thick material with a fourth covering every joint. While the version of the suit worn inside the spacecraft weighed 35 pounds (16 kg), the outdoor version weighed 50 pounds (23 kg), not including the PLSS backpack.

The helmet had two visors. The inner one acted like normal sunglasses to combat glare. The outer one was electroplated with gold to reflect the blinding rays of the Sun.

The gloves presented a particular problem as they ballooned due to the internal pressure of eight pounds per square inch. This made any movement fatiguing. Also their fingernails scraped painfully against the inner lining.

## GETTING OUT THE DOOR

Before going outside, the LM had to be depressurized. This was a slow process as the expelled atmosphere had to be passed through a biological filter to prevent germs from Earth being released on the Moon. Then the door stuck, which Buzz Aldrin remembered well:

> *We tried to pull the door open and it wouldn't come open. We thought, "I wonder if we're really gonna get out or not?" It took an abnormal time for it to finally get to a point where we could pull on a fairly flimsy door. You don't want to rupture that door and leave yourself in a vacuum for the rest of the mission.*

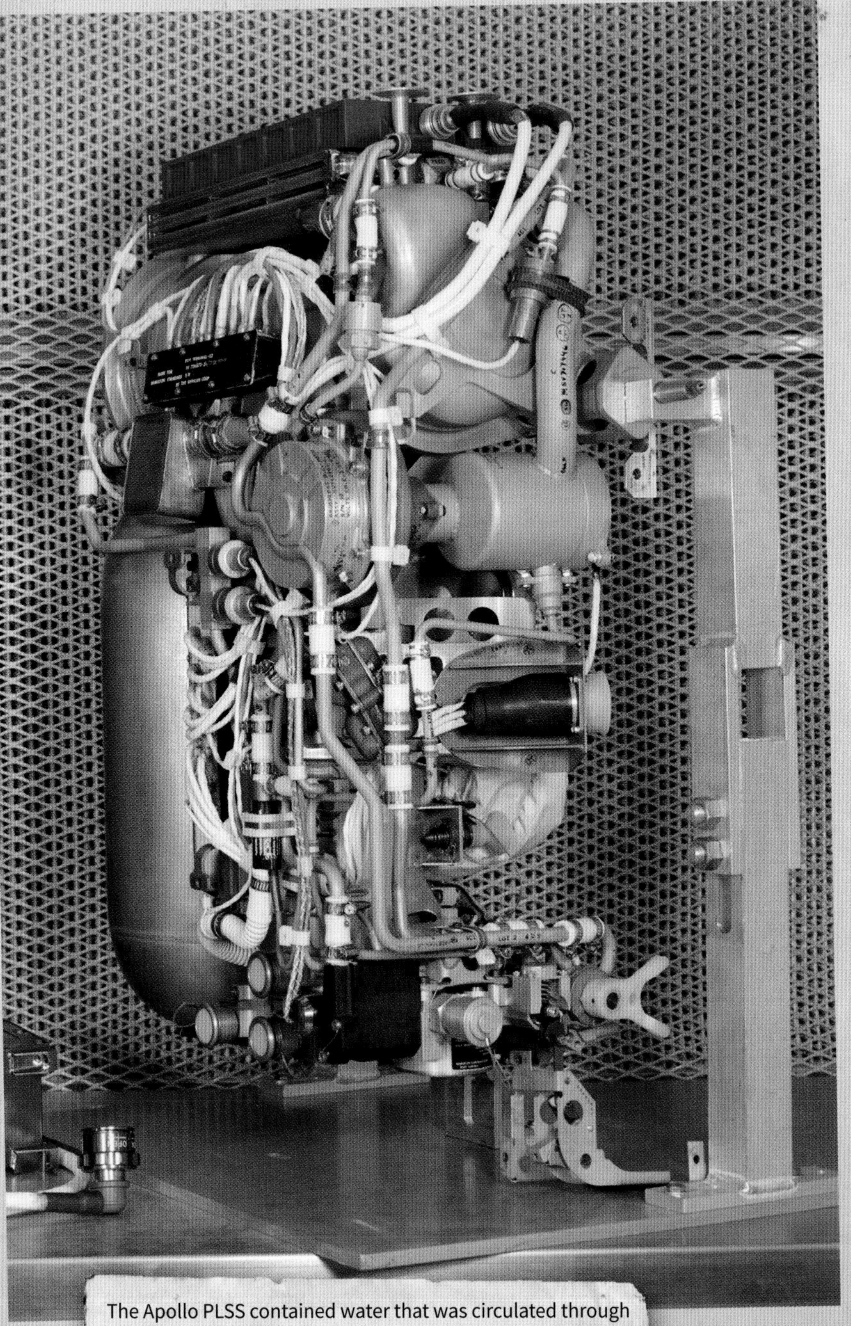

The Apollo PLSS contained water that was circulated through the astronaut's liquid-cooled undergarments, which cooled the air they breathed as well. It also supplied oxygen and carried a radio transceiver and camera. All these functions were regulated by a control panel on the astronaut's chest.

Back on Earth, Jan Armstrong thought the delay was because her husband was trying to figure out what he was going to say when he first stepped on the Moon. The hatch finally opened and Aldrin guided Armstrong as he crawled backward out of the *Eagle*'s narrow doorway.

"Neil, you're lined up nicely," said Aldrin. "Toward me a little bit, OK down, OK make it clear ... Here roll to the left. OK, now you're clear. You're lined up on the platform. Put your left foot to the right a little bit. OK, that's good. Roll left."

"That's OK? How am I doing?" said Armstrong.

"You're doing fine."

Simply getting out of the door took such concentration that Armstrong forgot to turn the handle that released the hatch of the Modular Equipment Storage Assembly (MESA) that contained everything they would need on the surface—including the TV camera that was to film the first step. He was already descending the ladder to the surface when Mission Control reminded him of his oversight. Armstrong then had to clamber back up again to turn the handle before resuming his descent.

"Houston, the MESA came down all right," Armstrong confirmed.

"This is Houston, roger. We copy," said CapCom Bruce McCandless. "And we're standing by for your TV."

# THE EVA SUIT

The Extravehicular Mobility Units worn by Armstrong and Aldrin weighed 48 pounds (22 kg), while its Portable Life Support System (PLSS) weighed another 57 pounds (26 kg). Next to the skin the astronaut wore long johns whose fabric contained a network of fine tubes carrying water to cool their bodies, which was circulated by pumps in their backpacks. This was covered with a layer of lightweight nylon.

Next came an airtight layer of Neoprene-coated nylon with bellows-like molded joints. This was pressured by oxygen from the backpack and covered with nylon restraint layers to prevent it ballooning. The water and oxygen flow were regulated by controls mounted on the chest.

Thermal insulation was provided by alternating layers of Kapton and glass-fiber cloth, several layers of Mylar and spacer material. These were aluminized to make them reflective. The whole thing was covered with a layer of Teflon-coated glass-fiber cloth to protect against puncturing by jagged rocks and micrometeorites. The suit came equipped with a diaper and bag to collect urine.

The gloves worn on the lunar surface were also multilayered and pressurized. They were molded from casts of the astronauts' hands. The thumb and fingertips were made of silicone rubber to provide sensitivity and some sensation of feel.

Their helmets were made from high-strength polycarbonate and attached to the spacesuit by a pressure-sealed neck ring. Unlike the earlier helmets used on the Mercury and Gemini missions which moved with the astronaut's head, the helmets used on the Moon walk were larger, so the astronaut's head moved freely within them. An outer visor protected their eyes against ultraviolet radiation.

Within the helmets they wore communications skullcaps with connectors to the radio transmitters and receivers in their backpacks, which also sent back medical data from the body sensors they wore under their suits. A bag of water was draped around the astronaut's neck with a straw for drinking.

The lunar boots they wore were actually overshoes that fitted over the pressurized boots that were an integral part of the spacesuit. The outer layer was made from metal-woven fabric with a ribbed silicone rubber sole and a tongue area made from Teflon-coated glass-fiber cloth. The boot inner layers were made from Teflon-coated glass-fiber cloth followed by twenty-five alternating layers of Kapton film and glass-fiber cloth for thermal insulation.

The suits were tested in a vacuum chamber on Earth. One test went dangerously wrong when the hose pressurizing the suit became disconnected and the internal pressure dropped from 3.8 pounds per square inch to 0.1 psi in ten seconds.

"As I stumbled backwards, I could feel the saliva on my tongue starting to bubble just before I went unconscious and that's the last thing I remember," said test subject Jim LeBlanc.

"Essentially, he had no pressure on the outside of his body and that's a very unusual case to get," explained Cliff Hess, the supervising engineer. "There's very little in the space medical literature about what happens when you have that. There's a lot of conjecture, that your fluids will boil."

The chamber, which would normally take thirty minutes to repressurize, was blasted back to atmospheric pressure in eighty-seven seconds. Amazingly, LeBlanc survived suffering only an earache from his ordeal.

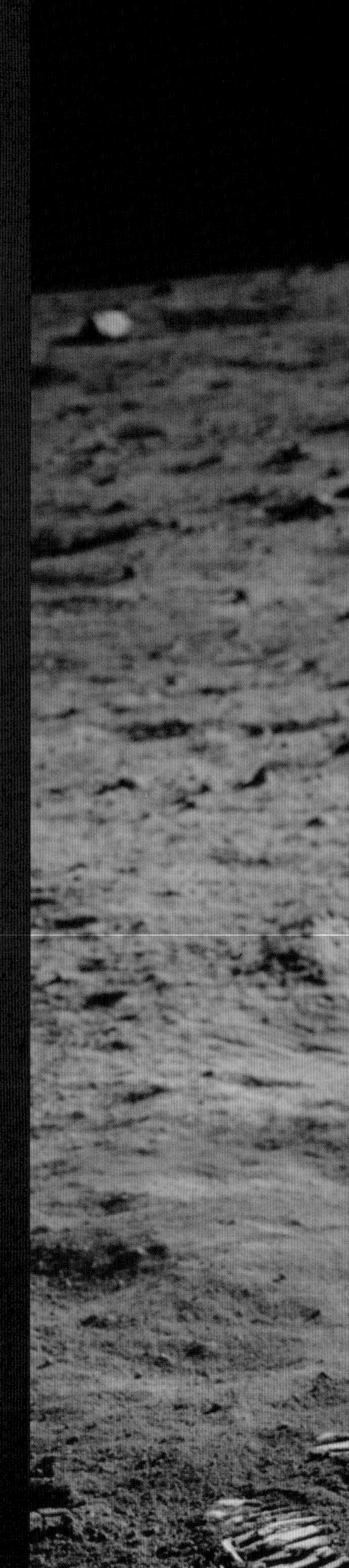

Buzz Aldrin stands on the surface of the Moon during the Apollo 11 moonwalk. Neil Armstrong took this photograph with a 70mm lunar surface camera. The Lunar Module and Armstrong can be seen reflected in Aldrin's visor.

## COMING DOWN THE LADDER

There had been problems with TV reception at NASA's Goldstone Observatory in California, so it was patched via the Honeysuckle Creek tracking station near Canberra in Australia. The audio was still coming into Goldstone and they went through radio checks.

Everyone at Mission Control began to cheer and clap when a picture appeared on the big screen.

"Roger. We're getting a picture on the TV," said McCandless.

"You got a good picture, huh?" asked Aldrin.

"There's a great deal of contrast in it, and currently it's upside down on our monitor, but we can make out a fair amount of detail," McCandless replied.

"Okay. Will you verify the position—the opening I ought to have on the camera?" asked Aldrin.

"Stand by ... Okay. Neil, we can see you coming down the ladder now," said McCandless. He then gave Aldrin instructions for altering the shutter speed on the sequence camera as coming down the ladder facing away from the Sun left his face in almost total darkness.

Then Armstrong told Houston: "I'm at the foot of the ladder. The LM footpads are only depressed in the surface about one or two inches. Although the surface appears to be very, very fine-grained, as you get close to it. It's almost like a powder. Now and then, it's very fine. I'm going to step off the LM now."

It was estimated that 600 million people—one-fifth of the world's population—were watching or listening to what happened next. The only person who could not see it was Mike Collins, orbiting above in *Columbia*.

This is the television image that was transmitted to the world on July 20, as Armstrong egressed the ladder to the lunar surface. The black bar running through the center of the photograph is an anomaly in the TV Ground Data System at Goldstone Tracking Station.

## THE GIANT LEAP

The last rung of the ladder was 3 feet 6 inches (1.06 m) above the surface. Armstrong dropped from it onto the *Eagle*'s footpad, then yanked himself back onto the ladder to make sure that the two astronauts could get back onto the ladder after their EVA.

Then as Armstrong's foot finally touched the surface of the Moon, he said: "That's one small step for (a) man. One giant leap for mankind."

Armstrong was adamant that he said the "a" but NASA transcribers said they could not find it on the tape. It may have been obscured by static.

"The 'a' was intended. I thought I said it. I can't hear it when I listen on the radio reception here on Earth, so I'll be happy if you just put it in parenthesis," said Armstrong at a press conference in 1982.

NASA denied that Armstrong's historic words were written for him. Indeed, most were shocked that he said anything at all as Armstrong was known to be a man of few words. However when he set foot on the Moon, he became positively garrulous telling Houston:

*The surface is fine and powdery, I can ... I can pick it up loosely with my toe. It does adhere in fine layers like powdered charcoal to the sole and sides of my boots. I only go in a small fraction of an inch, maybe an eighth of an inch, but I can see the footprints of my boots and the treads in the fine, sandy particles.*

*There seems to be no difficulty in moving around as we suspected. It's even perhaps easier than the simulations at one-sixth g that we performed in the various simulations on the ground. It's actually no trouble to walk around. Okay. The descent engine did not leave a crater of any size. It has about one foot clearance on the ground. We're essentially on a very level place here. I can see some evidence of rays emanating from the descent engine, but a very insignificant amount.*

## MAGNIFICENT DESOLATION

Aldrin followed Armstrong out of the LM, remembering not to close the door behind him as the designers had neglected to put a handle on the outside. In the one-sixth gravity, the two of them bounced about like children. Buzz Aldrin said:

*The surface of the Moon was like fine talcum powder … When you put your foot down in the powder, the boot-print preserved itself exquisitely. When I would take a step, a little semicircle of dust would spray out before me. It was odd, because the dust didn't behave at all the way it behaves here on Earth. On Earth, you're sometimes dealing with puffy dust, sometimes with sand. On the Moon, what you're dealing with is this powdery dust traveling through no air at all, so the dust is kicked up, and then it all falls at the same time in a perfect semicircle.*

Aldrin talked of the "magnificent desolation" of the Moon. Both men remarked on the pronounced curvature of the horizon, showing they were standing on an orb. It also appeared close.

*[The Moon has] a stark beauty all its own, like much of the high desert of the United States. It's different but it's very pretty down here.*

Neil Armstrong

## CAUGHT BETWEEN LIGHT AND SHADE

Both astronauts remarked on the stark contrast between light and shade. Aldrin said: "Stepping out of the LM's shadow was a shock. One moment I was in total darkness, the next in the Sun's hot floodlight. From the ladder I had seen all the sunlit moonscape beyond our shadow, but with no atmosphere, there was absolutely no refracted light around me. I stuck my hand out past the shadow's edge into the Sun, and it was like punching through a barrier into another dimension."

The brilliant sunlight also caused problems.

"The light was sometimes annoying, because when it struck our helmets from a side angle it would enter the faceplate and make a glare that reflected all over it," Armstrong said. "As we penetrated a shadow we

Buzz Aldrin egressing the Lunar Module (LM) during the Apollo 11 extravehicular activity (EVA) on the Moon.

would get a reflection of our own face, which would obscure everything else. Once when my face went into shadow it took maybe twenty seconds before my pupils dilated out again and I could see details."

At one point Armstrong discovered that if he stuck up his thumb and closed one eye, he could blot out the Earth. "I didn't feel like a giant," he said. "I felt very, very small."

In their EVA suits they felt very cut off from the environment. There was no sense of touch, no smell, no sound, and even their vision was tinted by the visor. The only thing they could hear was the hum of the pumps in their backpacks, circulating fluid.

"You don't hear any amplified breathing inside your mask," said Aldrin. "That's a Hollywood contrivance. The name of the game on the Moon is stay cool and don't exert too much so you're never out of breath."

## MEN FROM PLANET EARTH

The astronauts had to-do lists sewn on their sleeves. The first task was to show the TV cameras the commemorative plaque bolted to the bottom of the *Eagle*. It was attached to the descent stage, so it would be left behind on the surface of the Moon after they departed, and had the following message:

> HERE MEN FROM THE PLANET EARTH
> FIRST SET FOOT UPON THE MOON
>
> JULY 1969 A.D.
>
> WE CAME IN PEACE FOR ALL MANKIND

Under it were engraved the signatures of Armstrong, Aldrin, Collins, and President Nixon. Nixon had wanted the words "Under God" added after "Peace," but NASA had enough problems with religious issues already and left it off.

A priority for Armstrong was supposed to have been to collect a rock sample in case the Moonwalk was suddenly aborted. Instead he became engrossed in taking pictures with the stills camera that the Swedish manufacturer Hasselblad had specially adapted for the voyage. It had the first motorized winding system and an enlarged magazine. In some 2 hours 30 minutes on the surface, they would take 857 black-and-white and 550 color photographs.

Strangely there were only four photographs of Armstrong on the Moon, each showing him as a small figure in the lunar landscape. The only close-up was one he took himself, reflected in Aldrin's visor.

After handing the camera over to Aldrin, Armstrong set about collecting soil samples. "It's a very soft surface but here and there where I plug with the contingency sample collector, I run into a very hard surface but it appears to be very cohesive material of the same sort," he told Houston. "I'll try to get a rock in here. Here's a couple."

The sample was put in a Teflon bag, which he tried to put into a pocket on the leg of his spacesuit. However, as the spacesuit was pressurized rigidly, this proved impossible and Aldrin had to help.

Armstrong then put the television camera on a tripod to capture the panorama, while Aldrin set up the equipment to take sub-surface samples. Then he deployed the solar wind collector. This was essentially a stand holding a sheet of aluminum that trapped particles and could be rolled up and taken back to Earth for analysis. As the Moon has little atmosphere or magnetic field, particles escaping the outer layers of the Sun reach the surface.

Buzz Aldrin carrying the EASEP out to the deployment site a short way south of the spacecraft.

## SALUTING THE FLAG

Then came the moment to unfurl the American flag. This proved harder than anticipated as under just a few inches of dust was impenetrable rock. Armstrong had to pat together a small mound of dust to support the flag and spent the rest of the EVA terrified that it would topple over.

The other problem with the flag was that, with no atmosphere, there could be no wind to lift it from hanging limp against the flagpole. Anticipating this, a flexible aluminum tube was inserted in a hem sewn along the top of the flag. The astronauts then found that if they did not extend the tube completely, the cloth appeared rippled as if being blown by the wind.

As Armstrong and Aldrin saluted the flag, everyone in Mission Control stood and cheered. It meant a lot to the astronauts.

"Being able to salute that flag was one of the more humble yet proud experiences I've ever had," Aldrin said. "To be able to look at the American flag and know how much so many people had put of themselves and their work into getting it where it was. We sensed—we really did—this almost mystical identification with all the people in the world at that instant."

## WHITE HOUSE CALL

After the ceremony with the flag was complete, Houston patched through a call from President Nixon:

## The Stars and Stripes on the Moon

The United Nations Treaty on Outer Space prohibits territorial claims, so the planting of the American flag was a matter of controversy. Congress then took up the issue, debating whether American astronauts should plant the flag of the UN or the stars and stripes on the Moon, while some maintained that it should be a Christian flag. The matter was on the floor just as the appropriations bill for the following year was under consideration.

NASA depended on this bill for its funding, so on June 10, 1969, it informed Congress that the US flag would be raised on the lunar surface. That same day the appropriations bill was passed with an amendment saying that:

> *The flag of the United States, and no other flag, shall be implanted or otherwise placed on the surface of the Moon, or on the surface of any planet, by members of the crew of any spacecraft ... as part of any mission ... the funds for which are provided entirely by the Government of the United States.*

In deference to the UN Treaty, the amendment concluded by saying:

> *This act is intended as a symbolic gesture of national pride in achievement and is not to be construed as a declaration of national appropriation by claim of sovereignty.*

As the decision to carry the flag had been taken so late, it had to be strapped to the leg of the LM in a steel-and-Thermoflex casing to protect it from the blast of the descent engine during touchdown.

*Hello, Neil and Buzz, I am talking to you by telephone from the Oval Room at the White House, and this certainly has to be the most historic telephone call ever made. I just can't tell you how proud we all are of what you've done. For every American, this has to be the proudest day of our lives, and for people all over the world, I am sure they, too, join with America in recognizing what an immense feat this is. Because of what you have done, the heavens have become a part of man's world, and as you talk to us from the Sea of Tranquility it inspires us to redouble our efforts to bring peace and tranquility to Earth. For one priceless moment, in the whole history of man, all the people on this Earth are truly one. One in their pride in what you have done, and one in our prayers that you will return safely to Earth.*

"Thank you, Mr. President," Armstrong said. "It's been a great honor and privilege for us to be here representing not only the United States but men of peace of all nations, and with interest and curiosity, and men with a vision for the future. It's an honor for us to be able to participate here today."

## MOONWALKING

Armstrong and Aldrin went back to collecting geological samples. This again proved harder than expected. One task was to get a core sample 18 inches (45 cm) deep. But even hammering the collection tube until it was dented would only drive it in 5 inches (12 cm).

However, moving around on the Moon was easy, despite the stiff joints that restricted the movement of the knees. Instead the astronauts used their ankles and toes to bounce along. This gait became known as the "kangaroo hop." Stopping or changing direction took practice, but the Moon's one-sixth gravity meant that the men's weight combined with that of the suit and PLSS of 360 pounds (163 kg) was effectively just 60 pounds (27 kg).

"You can actually just fall over on your face like a dead man, right down to the surface, and push yourself back up," said Armstrong. Aldrin also found the experience enjoyable:

*The Moon was a very natural and very pleasant environment in which to work. It had many of the advantages of zero-gravity, but it was in a sense less lonesome than zero-g, where you always have to pay attention to securing attachment points to give you some means of leverage. In one-sixth gravity, on the Moon, you had a distinct feeling of being somewhere, and you had a constant, though at many times ill-defined, sense of direction and force … As we deployed our experiments on the surface we had to jettison things like lanyards, retaining fasteners, etc., and some of these we tossed away. The objects would go away with a slow, lazy motion. If anyone tried to throw a baseball back and forth in that atmosphere he would have difficulty, at first, acclimatizing himself to that slow, lazy trajectory; but I believe he could adapt to it quite readily.*

Armstrong (left) and Aldrin planting the American flag on the surface of the Moon.

## NEIL ARMSTRONG'S SEISMIC ACTIVITY

While Armstrong went off to examine a nearby crater which contained some unusual-looking rocks, Aldrin set up the rest of the Early Apollo Surface Experiment Package. Its Passive Seismic Experiment Package (PSEP) was composed of four solar-powered seismometers—three long-period seismometers and one short-period—designed to measure the effects of meteorite impacts and moonquakes. The data was then transmitted back to Earth.

The other experiment in the package was the Laser Ranging Retro Reflector. This reflected back a laser beam fired from the Earth to measure the exact distance from the Earth to the Moon.

The amount of science that could be done on the first lunar landing was restricted by the ten-minute rule. General Samuel C. Phillips, director of the Apollo Manned Lunar Landing Program, ruled that no science was to be done on the first mission that would not produce results in ten minutes—in case the mission was aborted soon after landing.

Indeed, the PSEP which was set up 30 feet (9.1 m) from the LM did pick up something that night while the astronauts were having a pre-launch sleep. Though only a small effect, it did not seem to come from a meteorite strike at a distance. Rather it was closer to home. It was the flight surgeon who realized that the seismometer registered at the exact moment Armstrong turned over in his sleep.

Once the experiments had been laid out, Aldrin continued taking pictures with the Hasselblad, including one of his bootprints in the lunar dust. He also used the Gold Surface Close-Up 3D camera, designed to take stereoscopic pictures of the details of the lunar surface. But Armstrong said that there were problems:

> *The one thing that gave us more trouble than we expected was the TV cable. I kept getting my feet tangled up in it … Fortunately, Buzz was able to notice this and keep me untangled … he was able to tell me which way to move my foot to keep out of trouble. We knew this might be a problem from our simulations, but there just was no way that we could avoid crossing back and forth across that cable.*

Aldrin deploys the Passive Seismic Experiment (PSE).

## TIME TO GO HOME

Then the EVA was over and Mission Control told the two astronauts to return to the LM.

"You have approximately three minutes until you must commence your EVA termination activities," said CapCom McCandless. "Neil, this is Houston. Did you get the Hasselblad magazine?"

"Yes, I did," Armstrong replied. "And we got about, I'd say, 20 pounds (9 kg) of carefully selected, if not documented, samples."

In fact he had collected 48 pounds (22 kg) of rock and dust which were vacuum-packed in two-foot aluminum cargo boxes. Armstrong's heart rate hit 160 as he loaded the samples and equipment back into the LM. Aldrin's had only hit a maximum of 125 during the EVA.

They were supposed to have carried out another ceremony on the Moon. In February 1969, NASA had set up a Committee on Symbolic Articles to decide what items should be left behind by the first men on the Moon. But there was no time. Once back in the

Buzz Aldrin has just deployed the Solar Wind Collector (SWC). The scratch marks seen on the lower right corner were caused when Neil Armstrong pulled out the TV cable, scraping it along the ground.

LM, Aldrin pulled the collection from his pocket and dropped them out of the door.

There were patches honoring the dead from Apollo 1—Grissom, White, and Chaffee; Soviet medals for Gagarin and Komarov; a silicon wafer etched with goodwill messages from seventy-three heads of state and other key figures in the project that could be read by a sixty-power microscope; and a golden olive branch symbolizing peace. These items had been authenticated in the CSM on the way to the Moon so Collins would not be left out of the proceedings.

## LIGHTENING THE LOAD

Back in the *Eagle*, Armstrong and Aldrin plugged back into the LM's life support systems, so they could take off their backpacks and throw them out. These were followed by their outer suits, boots, gloves, and visors, along with used filters and food packs, cameras, and urine bags to lighten the load.

"We observe your equipment jettison on TV and the passive seismic experiment reported shocks when each PLSS hit the surface. Over," said McCandless.

"You can't get away with anything anymore, can you?" Armstrong replied.

They drank fruit punch and ate Vienna sausages while answering questions from the scientists and engineers at Mission Control. One query was about their perception of temperature outside the LM. Armstrong said he had no idea, guessing it was somewhere between zero and 100°C (32 – 212°F).

"I really wasn't aware of any temperatures inside the suit," he said. "And at no time could I detect any temperature penetrating the insulated gloves as I touched things—the LM itself, things in the shadow, things in the sunlight, the tools, the flagpole, the TV camera, the rocks that I held."

There was a surprise in store for them when they repressurized the LM and removed their helmets. There was a strange smell which Armstrong likened to wet ashes in a fireplace, guessing that this was coming from the moondust they had brought back into the cabin.

By then the two astronauts had been at work for over twenty-two hours, so they tidied up the cockpit as best as they could. They decided to sleep with their helmets and gloves on to prevent them breathing in the dust. Also it would be quieter. It was surprisingly noisy in the LM with switches and a pump going on and off.

It was also very cold. The only way they could have warmed the ship was to raise the window shades, which would have ruled out sleeping. Aldrin took the hammock, while Armstrong lay on the floor with his head under the telescope so the Earth appeared like a huge unblinking eye glaring down on him.

Armstrong had just three hours sleep, Aldrin had four. During the night, the Soviet probe Lunar 15 had approached its lunar landing site 500 miles (805 km) to the south. It was going to scoop up samples and return them to the Earth, taking the edge off Apollo 11's achievement. Instead it crashed, leaving America's lead in the space race unassailable.

> *If I thought about the odds at all, we'd never get to the pad.*
>
> Flight director Chris Kraft

## LONELIEST MAN SINCE ADAM

Meanwhile Collins kept orbiting above. According to NASA's press office: "Not since Adam has any human known such solitude as Mike Collins is experiencing during this forty-seven minutes of each lunar revolution when he's behind the Moon with no one to talk to except his tape recorder aboard *Columbia*."

When he returned to Earth, Collins received a letter from Charles Lindbergh, who had some earlier experience of flying solo. But for Collins it was no more unnerving than when he flew F-86s across Greenland in winter.

However, Collins did have problems of his own. The coolant temperature control in *Columbia* was malfunctioning. To solve the problem he was supposed to go through an elaborate procedure. Instead, he switched the control from automatic to manual, then back to automatic which seemed to solve the problem.

The other task he was set was to locate the LM as the data sent back from the *Eagle* was inconsistent. Nor did it tally with that from the tracking stations on Earth. After hours of searching, he gave up.

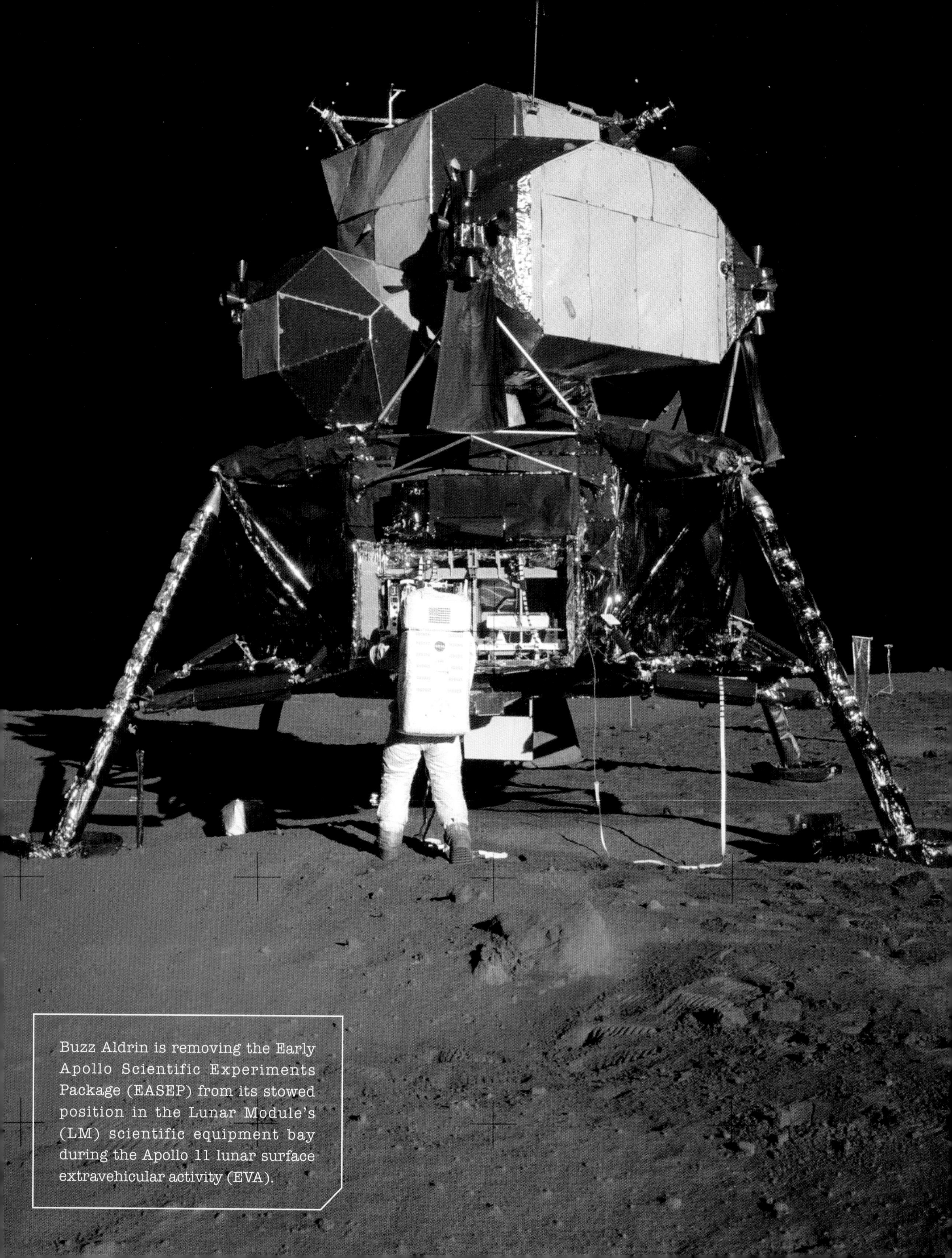

Buzz Aldrin is removing the Early Apollo Scientific Experiments Package (EASEP) from its stowed position in the Lunar Module's (LM) scientific equipment bay during the Apollo 11 lunar surface extravehicular activity (EVA).

# CHAPTER 16

# LUNAR LIFTOFF

After all too brief a sleep, Armstrong and Aldrin were woken by Houston. They had a freeze-dried breakfast of cookies, peaches, and bacon to raise their blood sugar level. They needed to be alert for what would be the most fraught procedure in the entire voyage—lunar liftoff. Grumman lunar module program director Joseph Gavin said:

> *The critical thing was the take-off, because you had a limited time, you had to punch the button, and everything had to work. The ascent engine had to ignite. The explosive bolts had to explode. The guillotine had to cut the connections, and then it had to fly up. And this is something we never saw happen until the last mission. So it was all, well, hearsay. It's something we never could test for, because the conditions couldn't be duplicated on Earth.*

Almost everything on Apollo 11 had a backup in case of failure. But due to weight restrictions, no backup could be built into the ascent engine. In testing too, it had proved worryingly error-prone.

## A MARKED MAN

The liftoff procedure was simple. All Armstrong had to do was flip the switches marked "Abort Stage" and "Engine Arm," then Aldrin had to press the button marked "Proceed" and the rocket would ignite. The engine was so powerful and the *Eagle* so light that it would accelerate to 3,000 mph (4,828 kph) in two minutes. But what happened if it did not ignite? There was nothing Michael Collins in the CM could do to help and he knew it:

> *I have skimmed the Greenland ice cap in December and the Mexican border in August; I have circled the Earth forty-four times aboard Gemini 10. But I have never sweated on any flight like I am sweating on the LM now. My secret terror for the last six months has been leaving them on the Moon and returning to Earth alone. If they fail to rise from the surface, or crash back into it, I am not going to commit suicide; I am coming home, forthwith, but I will be a marked man for life and I know it ... I would do everything I could to help them. But they know and I know, and Mission Control knows, that there are certain conditions and malfunctions where I just simply light the motor and come home without them.*

He had been trained in eighteen emergency rendezvous procedures, but in none of them could he drop down below 50,000 feet (15,240 m) as the height of some of the mountains on the Moon were unknown. Some were certainly over 30,000 feet (9,144 m).

## SOARING ABOVE THE CRATER FIELDS

CapCom Ron Evans gave the astronauts the go-ahead for takeoff.

"Roger, understand we're number one on the runway," said Aldrin. He then turned off the rendezvous radar to prevent the computer overload that had caused problems on the descent. While Armstrong tested the thrusters, Aldrin fired two small charges that would open the helium tanks and pressurize the system. But the pressure gauge on the *Eagle*'s dashboard did not flicker. He tried again. Still there was no response. But there was nothing to be done. The countdown continued in the hope that the problem was merely a stuck gauge.

With a minute to go before liftoff, Aldrin armed the explosive bolts that would free the ascent stage. Then they hit another problem, as Aldrin explained:

> *On the corner of the floor I saw a small black object and I immediately recognized what it was. It was the end of a circuit breaker that had broken off. So I looked up to see these rows of circuit breakers, see which one had broken off, and it was the one that armed the engine to ignite … So I used a pen that we had and I pushed it in, and it stayed in. But there wasn't any way to pull it back out again, if you needed to do that.*

*They know and I know, and Mission Control knows, that there are certain conditions and malfunctions where I just simply light the motor and come home without them.*

Michael Collins, Apollo 11
Command Module Pilot

The explosive bolts fired, freeing the *Eagle* from the descent stage and its tanks. The ascent engine ignited, producing 3,500 pounds of force and the ascent stage began to rise. The exhaust churned up a cloud of dust and blew over the American flag.

"Our liftoff was powerful. Nothing we'd done in the simulators has prepared us for this amazing swoop upward in the weak lunar gravity," said Aldrin. "Within seconds, we had pitched forward at a sharp forty-five degrees and were soaring above the crater fields."

The rocket fired for 7 minutes 45 seconds. The ride was far from smooth as the computer navigation system fired the thrusters to keep it on course. These wobbled the flimsy craft as they could not afford the fuel to fire retro-rockets to stabilize it.

## RENDEZVOUS IN SPACE

The rocket accelerated the ascent stage to 3,775 mph (6,075 kph), putting it in an elliptical orbit of 53.7 by 10.8 miles (86.4 by 17.4 km), while *Columbia* was somewhat higher at 72.7 by 65.4 miles (117 by 105.3 km). The engine was then fired again for forty-five seconds, raising it to rendezvous with the mother ship's orbit.

After two orbits of the Moon, the two ships lined up for docking. Collins was relieved to see the LM approaching, describing it as the mission's happiest moment. They were in darkness on the far side of the Moon when the two craft aligned.

Armstrong altered the flight plan slightly to make a more direct path to *Columbia*. After they made contact an abnormal oscillation was set up in the yaw axis. Collins said: "All hell broke loose." He feared that they might have to separate and make a second attempt at docking. But it did not prove necessary.

"I instantly took action to correct the angle, so did Neil in the *Eagle*," said Collins. "We heard a loud bang, which is characteristic of those twelve big latches slamming home. And lo and behold, we docked and it was all over."

Collins then dismantled the docking mechanism to open the tunnel between the two ships, while Armstrong and Aldrin vacuumed up as much moondust as they could to prevent it getting into the CM. Nobody knew what effect it might have if it was inhaled or got on the skin.

The hatch was opened and the three astronauts were reunited and there was cheering and laughing.

"Finally they got back into the command module, and I grabbed Buzz by both ears and I was going to kiss him on the forehead," said Collins. He checked himself and grabbed his hand, then Armstrong's.

The three transferred their precious cargo of lunar samples across to *Columbia*. The hatch was closed and the LM, now stuffed with anything they would not need for the trip home, was jettisoned. Later it crashed into the Moon. The impact registered on the seismometers they had left on the surface.

*The motor on the Lunar Module was one motor, and if something went wrong with it, they were dead men. There was no other way for them to leave.*

Michael Collins, Apollo 11
Command Module Pilot

## OUR ALIEN FRIENDS

*Columbia* was on the far side of the Moon when the rocket on the SM burned for 2 minutes 30 seconds boosting its speed to 5,300 mph (8,530 kph)—lunar escape velocity. It then began its two-and-a-half-day journey home.

The trip was uneventful, except for the strange noises that Houston heard over the audio from the spacecraft. "You sure you don't have anybody else in there with you?" asked CapCom Duke. "We had some strange noises coming down on the downlink, and it sounded like you had some friends up there."

There was some laughter. Later, at a press conference, the three astronauts admitted that they had made a tape of what they imagined friendly aliens to sound like which they played back as a prank.

## BROADCASTING TO THE WORLD

During trans-Earth injection, the astronauts made more TV broadcasts. Armstrong took the opportunity to show the viewers the boxes containing the lunar samples sealed in the vacuum of the surface of the Moon.

"These boxes include the samples of the various types of rock, the ground mass of the soil, the sand and silt and the particle collector for the solar wind experiment and the core tubes that took depth samples of the lunar surface," he explained.

The LM *Eagle* approaches the CSM *Columbia* for docking, with Earthrise in the background, July 21, 1969.

Once they had splashed down and the capsule was onboard the rescue ship, the samples would be sent to the Lunar Receiving Laboratory.

"We know there's a lot of scientists from a number of countries standing by to see the lunar samples," he said.

Aldrin showed the food they were eating onboard that day—shrimp cocktail and salmon salad.

"Another early development was the use of bite-size food," he said. "These bite-sized objects were designed to remove the problem of having so many crumbs floating around in the cabin, so they designed a particular size that would be able to go into the mouth all at once."

However, they had discovered that it was possible to eat food similar to that consumed on Earth.

"As a matter of fact, on this flight we've carried along pieces of bread, and along with the bread we have a ham spread. And I'll show you, I hope, how easy it is to spread some ham while I'm in zero-g."

Collins filled a spoon with water, then turned it upside down, showing that the water stayed in the spoon whichever way you turned it due to surface tension. Then he tossed some up and caught the flying globules in his mouth.

The LM *Eagle* from the rim of Little West Crater. With the sun behind him, Neil Armstrong's shadow is cast on the lunar surface as he takes the picture. One of the few images showing Armstrong on the Moon.

## GIANTS OF SCIENCE

In the last television broadcast to the watching millions back on Earth, Neil Armstrong put the trip in its historical context:

> *A hundred years ago, Jules Verne wrote a book about a voyage to the Moon. His spaceship, Columbia, took off from Florida and landed in the Pacific Ocean after completing a trip to the Moon. It seems appropriate to us to share with you some of the reflections of the crew as the modern-day Columbia completes its rendezvous with the planet Earth and the same Pacific Ocean tomorrow.*

Michael Collins paid tribute to the 400,000 men and women who had made their trip to the Moon possible. He told the TV audience:

> *This trip of ours to the Moon may have looked, to you simple or easy. I'd like to assure you that has not been the case... All this is possible only through the blood, sweat, and tears of a number of people ... This operation is somewhat like the periscope of a submarine. All you see is the three of us, but beneath the surface are thousands and thousands of others, and to all of those, I would like to say, "Thank you very much."*

Buzz Aldrin continued the broadcast, looking at the wider implications of the mission:

> *This has been far more than three men on a mission to the Moon; more, still, than the efforts of a government and industry team; more, even, than the efforts of one nation. We feel that this stands as a symbol of the insatiable curiosity of all mankind to explore the unknown ... We've been pleased with the emblem of our flight, the eagle carrying an olive branch, bringing the universal symbol of peace from the planet Earth to the Moon.*

Finally Neil Armstrong expressed his gratitude to the people of America and paid respect to those who had gone before:

> *The responsibility for this flight lies first with history and with the giants of science who have preceded this effort; next with the American people, who have, through their will, indicated their desire; ... We would like to give special thanks to all those Americans who built the spacecraft; who did the construction, design, the tests, and put their hearts and all their abilities into those craft. To those people tonight, we give a special thank you, and to all the other people that are listening and watching tonight, God bless you. Good night from Apollo 11.*

# CHAPTER 17

# COMING BACK DOWN TO EARTH

The SM's rocket fired again to direct the CM at the US Navy's recovery team comprising nine ships and fifty-four aircraft. NASA pointed out that it would be appropriate for the aircraft carrier USS *John F. Kennedy* to lead the operation. The Nixon White House sent USS *Hornet* instead.

Collins then shut down the SM. Once the systems were disconnected, he fired the explosive bolts and jettisoned it. At 12,250 pounds (5,557 kg), the remaining craft weighed just 0.2 percent of that of the original vehicle at blastoff.

Before the radio blackout resulting from the capsule plunging into the atmosphere, Houston sent one more message: "Have a good trip ... and remember to come in BEF." That is, Blunt End Forward, the correct orientation for reentry, with the heat shield—a two-inch-thick honeycomb of carbon fiber and ablation compound—facing into the onrushing atmosphere.

The capsule hit the atmosphere at 26,000 mph (42,000 kph) over the Solomon Islands. It had to be maintained at an angle of between –5.5° and –7.5°. Tilted too far forward it would burn up; too far back, it would bounce out of the atmosphere and they did not have enough fuel or oxygen for a second try.

The crew were hit with a deceleration force of 6.3g, while the temperature of the heat shield soared to 5,000°F (2,760°C) and it burned away, taking the heat with it.

*You are literally on fire, your heat shield's on fire, and its fragments are streaming out behind you ... It's like being inside a gigantic light bulb.*

Michael Collins, Apollo 11 Command Module Pilot

The air around the capsule as it sped through the atmosphere ionized, cutting off any radio communication for four minutes and soaking the capsule in a spectacular display of colored lights.

"You are literally on fire, your heat shield's on fire, and its fragments are streaming out behind you," Collins said. "It's like being inside a gigantic light bulb."

## SPLASHDOWN

On the way through the atmosphere, the computer swiveled the flight path, slowing the craft, and migrating the g-force on the crew. Then at 23,000 feet (7,000 m), the drogue parachutes were deployed. The Navy's tracking planes tried to locate the capsule, as Recovery Systems Chief John C. Stonesifer recalled:

> *I recall on the Apollo missions that I was out on, I don't recall, ever, one of them landing far enough away from the ship that we weren't able to see it, which was pretty exciting. Except Apollo 11. But the first clue we always had was the sonic boom, because when Apollo came back in, it was coming pretty much over the landing area, almost straight down, and you knew when you got that "boom, boom," we'd always yell, "It's overhead." So then we began to really start looking for it.*

At 10,000 feet (3,000 m) with the capsule still traveling at 175 mph (282 kph), the three main chutes were released. When the parachutes came out of their packing they were still furled with a line around the shroud. Small explosive cutters severed this line and the orange-and-white chutes billowed.

Apollo 11 splashed down seven minutes after the drogue parachutes had deployed. There was a 7-foot (2 m) swell and the sodden parachutes turned the capsule upside down. The crewmen were then suspended from the ceiling. Learning from Apollo 8, the three astronauts had taken seasick tablets before reentry. Collins managed to deploy the flotation collar and airbags which righted the craft.

The three Apollo 11 crewmen await pickup by a helicopter from USS *Hornet*. The fourth man in the life raft is a United States Navy underwater diver. All four men are wearing biological isolation garments.

## RECOVERING *COLUMBIA*

Four of USS *Hornet*'s Sea King helicopters arrived and a team of divers secured the capsule with a rope. A Navy frogman in a biological isolation suit closed the valves for filtering the capsule's air, allowing the hatch to open. He passed in three biological isolation suits. They dressed quickly, inflated their Mae Wests, and joined him in a raft. There they scrubbed each other with iodine and bleach.

There was thought to be a danger that the astronauts might had brought back some pathogen with them that might contaminate the seas and the fisheries and, on land, destroy crops.

"The thought processes there were that the heat of reentry would possibly destroy anything that was coming back on the exterior of the spacecraft, which seemed logical," said Stonesifer. "Or if when it landed into the water, the dilution factor of the ocean was another backup to the heat of reentry. But the spacecraft vented in the air. So there wasn't anything we could do about that."

With the astronauts safely on the raft, the Navy diver tried to close the capsule's hatch to prevent any further contamination, but he could not lock it. Armstrong also had a go. It was eventually secured by Collins.

While opening the hatch and getting the astronauts out of the capsule risked contamination, it was necessary to get them out in case it sank like Mercury 2. Another fault in the quarantine procedures was that the biological isolation suits were not watertight. They were also impossibly hot and the faceplates fogged over. The three astronauts were hoisted up into helicopters and flown to USS *Hornet*.

"We had special procedures for bringing the astronauts onboard, decontaminating the deck on which they walked, and getting them into the mobile quarantine facility," said Stonesifer.

## MOBILE QUARANTINE FACILITY

Onboard the *Hornet*, there were tanks of Betadyne, a disinfectant which was sprayed on the capsule. The men and their capsule took an elevator below decks where a trailer awaited complete with biological filters. This was the Mobile Quarantine Facility (MQF). The cargo was removed from the capsule down a plastic tube by engineer John Hirasaki. He had now come in contact with moondust, so he went into quarantine for three weeks with the three astronauts and flight surgeon Dr. William Carpentier.

The fliers celebrated their return to Earth with a cold martini, a hot shower, and a much needed shave. Unused to gravity, they were quickly exhausted. Even so, Aldrin and Collins tried to resort to normality with an exercise routine.

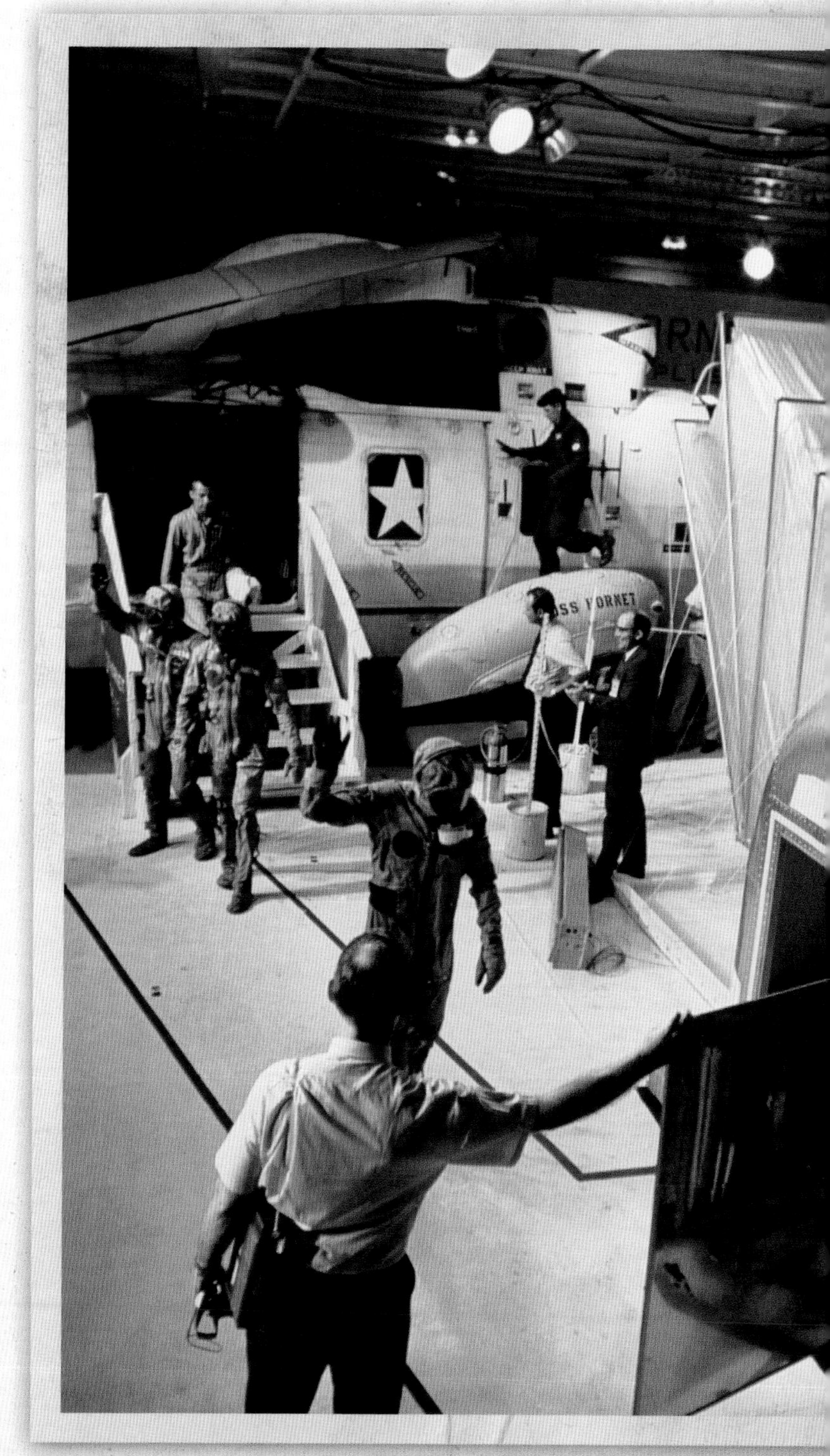


The Apollo 11 crewmen arrive aboard USS Hornet during recovery operations in the central Pacific.

On July 24, 1969, the Apollo 11 crew splashed down in the Pacific Ocean after achieving the first lunar landing. The crew underwent a 21-day quarantine to protect against the small possibility of lunar contagion. The photo shows left to right: Mission Commander Neil Armstrong, Command Module Pilot Michael Collins, and Lunar Module Pilot Buzz Aldrin in the Mobile Quarantine Facility.

HORNET
PLUS
THREE
APOLLO 11

US President Richard Nixon welcomes the Apollo 11 astronauts home aboard USS *Hornet*.

President Nixon was onboard USS *Hornet* to welcome the astronauts home. After a short speech, he chatted with them through the trailer window. The President described their eight-day journey as "the greatest week in the history of the world since the Creation." He went on to say:

> *As a result of what happened in this week, the world is bigger, infinitely … As a result of what you have done, the world has never been closer together before.*

There were celebrations in Houston and Huntsville where Wernher von Braun was carried shoulder high in a parade across town.

The *Hornet* put in to Pearl Harbor. From there the MQF was loaded onto a C-141 Starlifter cargo jet and taken to Ellington Air Force Base in Texas where the men were greeted by their wives.

The MQF was then lifted onto a flatbed truck and paraded through the streets of Houston to the Lunar Receiving Laboratory at the Manned Spacecraft Center. Before they were separated from their capsule, Collins sneaked back onboard and wrote next to the sextant:

> *Spacecraft 107—alias Apollo 11—alias Columbia.*
> *The best ship to come down the line.*
> *God Bless Her.*
> *Michael Collins, CMP.*

CMP stood for "Command Module Pilot."

## CAMEL DRIVERS RADIO CLUB OF KABUL

Although they remained in quarantine, the astronauts were allowed to roam the Lunar Receiving Laboratory which they shared with white mice who were monitored for ill-effects.

"Had the mice all sickened, jeez, I hate to think about it, we'd be in that building today," Collins said later.

They were joined by two cooks, a janitor, a photographer, a NASA public information officer, a researcher whose glove had ripped when he was handling moon rocks, and a photographic technician who had touched the dust on a film magazine. They filled the time by playing table tennis, pool, and gin rummy, jogging in the halls, riding an exercise bike, and watching movies. There was also a great deal of correspondence to handle.

"Duke Ellington was playing his new composition 'Moon Maiden' in the Rainbow Room and we are welcome there," said Collins. "Ditto the Steel Pier in Atlantic City, which offers the three of us $100,000 for a one-week stint. Another offers to name a hybrid orchid after me, and I sign a release authorizing a race horse to be called Michael Collins. May he orbit the track at unheard-of velocities, even in the mud. There are congratulatory messages from the Montgomery Police Department, the Catholic Daughters of America, the American Fighter Pilots Association, the Pope, the Peace Corps ... on and on it goes. There are honorary memberships in a host of organizations, my favorite being the Camel Drivers Radio Club of Kabul, Afghanistan."

## THE GIANT LEAP TOUR

Armstrong, Aldrin, and Collins were released from quarantine on August 10, but NASA secured a gentleman's agreement with the press to leave them and their families alone for another five weeks. Then a month later, on September 13, they would begin their "Giant Leap" tour.

Meanwhile with their families and a retinue from NASA, the three astronauts flew to New York. They were met at LaGuardia by Mayor John Lindsay, then helicoptered to Wall Street. From there they joined open-top limousines for a ticker-tape parade down the Canyon of Heroes in front of a crowd estimated at four million. Some threw the punch cards then used to program computers. Sometimes the stacks of cards did not break up and the limousines were left with dents.

The astronauts flew to Chicago for another parade, then on to Los Angeles for dinner with President Nixon and his wife. Other guests included California Governor Ronald Reagan, Chief Justice Warren Burger, forty-four state governors, Bob Hope, Red Skelton, and other celebrities. Vice President Spiro Agnew presented the astronauts with the Medal of Freedom. A fourth was given to flight controller Steve Bales in recognition of the other 400,000 men and women who contributed to the program. Medals had also been struck for the three astronauts who had lost their lives in the Apollo 1 fire and were presented to their wives.

In reply, Collins said he was proud to be a member of the Apollo team, while Aldrin also acknowledged the contribution of others, saying: "There are footprints on the Moon. Those footprints belong to each and every one of you, to all of mankind, and they are there because of the blood, the sweat, and the tears of millions of people." Armstrong added:

> *I was struck this morning in New York by a proudly-waved but uncarefully scribbled sign. It said: "Through you, we touched the Moon." It was our privilege today to touch America ... We hope and think that those people shared our belief that this is the beginning of a new era—the beginning of an era when man understands the universe around him, and the beginning of the era when man understands himself.*

## HOPE FOR THE FUTURE

That weekend, they returned to Houston for another parade, a barbecue at the Astrodome, and dinner with Frank Sinatra. The following Tuesday, the three astronauts flew to Washington to address a joint session of Congress. Each of them saw the Moon landing as the beginning of a great age of space exploration and looked to the future.

Aldrin said: "The first step on the Moon was a step toward our sister planets and ultimately toward the stars. 'A small step for a man,' was a statement of fact; 'a giant leap for mankind,' is a hope for the future. What this country does with the lessons of Apollo applies to domestic problems, and what we do in further space exploration programs will determine just how giant a leap we have taken."

---

*There are footprints on the moon. Those footprints belong to each and every one of you, to all of mankind ...*

Buzz Aldrin

---

Collins added: "During the flight of Apollo 11, in the constant sunlight between the Earth and the Moon, it was necessary for us to control the temperature of our spacecraft by a slow rotation not unlike that of a chicken on a barbecue spit. As we turned, the Earth and the Moon alternately appeared in our windows. We had our choice. We could look toward the Moon, toward Mars, toward our future in space ... or we could look back toward the Earth, our home, with its problems spawned over more than a millennium of human occupancy. We looked both ways. We saw both, and I think that is what our nation must do."

Concluding, Armstrong turned to astrology and spoke of the "Age of Aquarius" which it was broadly believed that the world was then entering. Indeed, it was a mantra of the burgeoning hippie movement. He then asked: "Where are we going?" His answer was more scientific, but full of optimism: "Man must understand his universe in order to understand his destiny ... Our successes in space lead us to hope that this strength can be used in the next decade in the solution of many of our planet's problems."

During the Giant Leap tour, the three astronauts visited twenty-three countries, and were guests of the Queen of England, Emperor Hirohito, and the Pope. That Christmas, Armstrong accompanied Bob Hope on his USO tour entertaining troops in Vietnam, Guam, Taiwan, Turkey, Italy, and Germany. In May he visited the Soviet Union where he gave Premier Aleksei Kosygin moon rock and a Soviet flag they had taken with them to the Moon. The Apollo program still had its part to play in the Cold War.

But the touring began to take its toll. The three astronauts and their wives began to resent being used as a propaganda tool. Collins was lucky in a way. Having not set foot on the Moon, he became the "forgotten man." Aldrin took umbrage at being the second man to set foot on the Moon, rather than the first, while Armstrong quickly discovered that, having spent his career as a military and test pilot, he was never going to be allowed to fly anything remotely dangerous any more. Certainly, he would never be allowed to travel into space again.

## FATE OF THE ASTRONAUTS

Neil Armstrong went on to become the deputy associate administrator for aeronautics at the Office of Advanced Research and Technology, where he sought to promote the new digital fly-by-wire technology in which computer systems took over from the mechanics of the old stick-and-rudder control of aircraft. But he remained in that job for barely a year. He went on to become professor of aerospace engineering at the University of Cincinnati.

In 1979, he resigned to become national spokesman for Chrysler automobile and he sat on the boards of various aerospace companies, including Thiokol who made the engines for the Space Shuttle. In 1986, he served on the presidential commission looking into the *Challenger* disaster after the Space Shuttle broke up during its ascent, killing all seven crew members. Essentially a private man, his unwanted fame put a great deal of pressure on him, leading eventually to the breakup of his marriage. He divorced and remarried in 1994.

In 2012, he died due to complications resulting from vascular bypass surgery. His cremated remains were buried at sea after a ceremony onboard USS *Philippine Sea*.

Buzz Aldrin also suffered from being in the limelight, finding it impossible to continue his career as a scientist and engineer after the Moonwalk. He fell into depression and alcoholism. Eventually he sought psychiatric help. He wrote about his ordeal in his memoir *Men From Earth* and served as chairman of the National Mental Health Association. He too divorced and remarried.

He wrote novels and appeared in a number of movies and TV shows, helped produce the computer game *Buzz Aldrin's Race Into Space* and promoted future manned space travel.

After the Giant Leap tour, Michael Collins retired from NASA and became Assistant Secretary of State for Public Affairs, then went on to become the first director of the Smithsonian's National Air and Space Museum. He wrote his autobiography *Carrying the Fire: An Astronaut's Journeys* and a number of books on spaceflight.

---

*There remains the task of deciding the next step. Will we press forward to explore the other planets or will we deny the opportunity of the future? To me the choice is clear. We must take the next step.*

George Mueller, Director, Manned Spacecraft Center

---

New York City welcomes Apollo 11 crewmen in a showering of ticker tape down Broadway and Park Avenue in a parade termed as the largest in the city's history. Pictured in the lead car, from right to left, are Neil Armstrong, Michael Collins, and Buzz Aldrin.

PART SIX

# THE CHAINS OF GRAVITY

*THE ROCKET WILL FREE MAN*
*FROM HIS REMAINING CHAINS—*
*THE CHAINS OF GRAVITY—*
*WHICH STILL TIE HIM TO THIS PLANET …*
*IT WILL OPEN TO HIM THE GATES OF HEAVEN.*

WERNHER VON BRAUN
NASA ROCKET ENGINEER

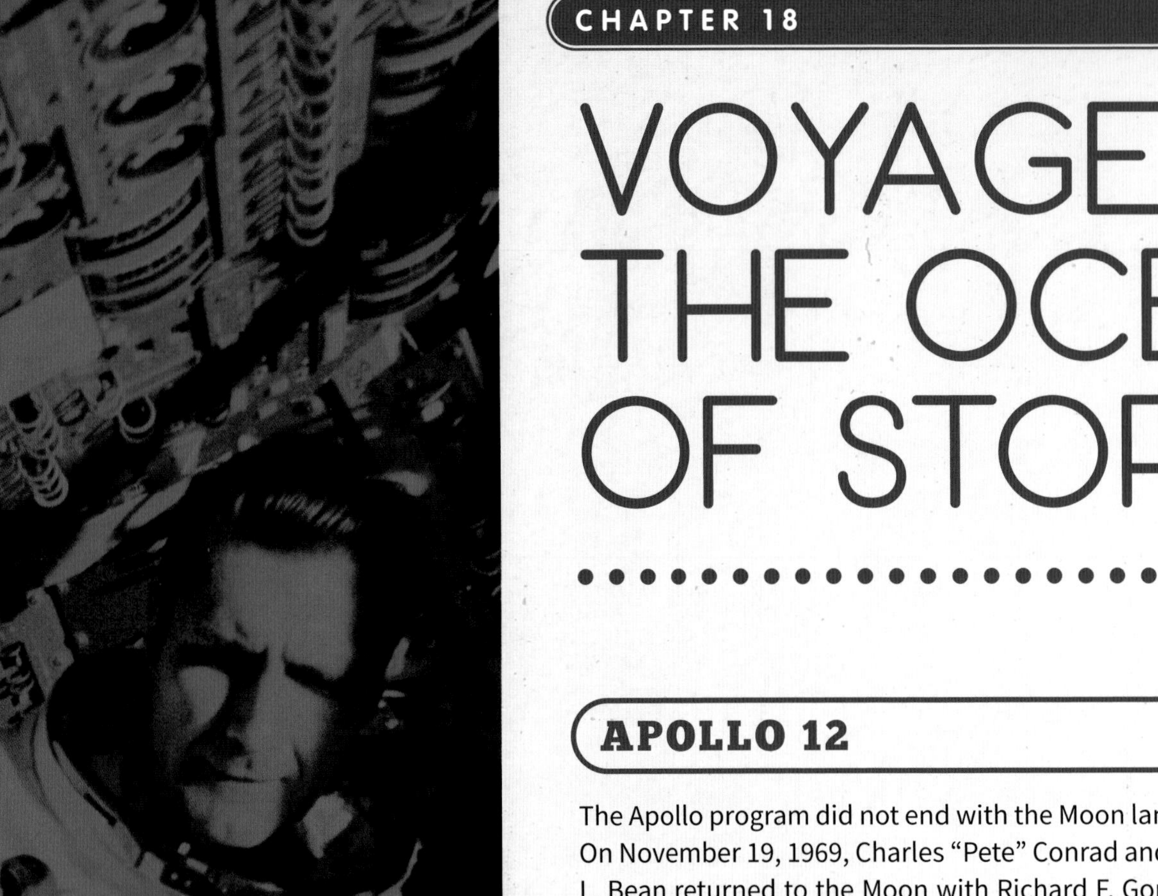

CHAPTER 18

# VOYAGE TO THE OCEAN OF STORMS

## APOLLO 12

The Apollo program did not end with the Moon landing. On November 19, 1969, Charles “Pete” Conrad and Alan L. Bean returned to the Moon with Richard F. Gordon orbiting in the CM. President Nixon attended the launch. He had stayed away from the launch of Apollo 11 and avoided any connection with it until Armstrong and Aldrin set foot on the Moon, in case the mission was a failure.

Moments after it blasted off, Apollo 12 was hit by two lightning strikes. These knocked out the fuel cells in the SM and much of the instrumentation. Telemetry was lost. However, a 24-year-old control engineer named John Aaron remembered that there was a seldom-used switch on the CM that put the systems on an auxiliary setting.

CapCom Gerry Griffin then relayed the instruction: “Apollo 12, Houston, try SCE to Auxiliary, over.” Ten miles up, Bean flipped the switch. Power instantly came back online and the computer systems started rebooting.

“I think we need to do a little more all-weather testing,” said Conrad.

As it was, nothing could stop the Saturn V which put Apollo 12 into Earth orbit. Checks were then run on the spacecraft. No significant problems were found, so the mission continued. The third stage was reignited for a second burn of 5 minutes 45 seconds, putting the craft into a lunar trajectory.

The CSM was separated, transposed, and reattached to the LM. The booster was then detached, but it failed to go into orbit around the Sun as planned. Conrad and Bean then moved through the tunnel from the CSM *Yankee Clipper* into the LM *Intrepid* to check out the systems. Again no damage from the lightning strike was detected. However, in Mission Control there were some fears that the explosive bolts that released the parachutes on reentry might have been damaged, dooming the astronauts when they returned to Earth. But as nothing could be done about it, it was decided not to tell the astronauts.

When Apollo 12 went behind the Moon, the engine was burned for six minutes, putting it into a lunar orbit of 69 by 195 miles (111 by 313 km). After two orbits, the engine was burned once more, putting it into a more circular orbit of 62 by 76 miles (100 by 122 km).

## LANDING IN THE OCEAN OF STORMS

Later that day Conrad and Bean entered the LM again and it separated from the CSM. On the fourteenth orbit, the LM's decent engine fired for twenty-nine seconds, lowering *Intrepid*'s descent orbit to 9 by 69 miles (14 by 111 km). It then landed in the Ocean of Storms, 535 feet (163 m) from its intended target—Surveyor 3, the third unmanned lander NASA sent to the Moon in 1967.

Apollo 12 carried a color TV camera, instead of the black-and-white camera carried by Apollo 11. Unfortunately, while setting it up, Bean inadvertently pointed it directly at the Sun, putting it out of action. The astronauts also took measurements of seismic activity on the Moon, the magnetic field, and the solar wind, as well as laying out the nuclear-powered Apollo Lunar Surface Experiments Package (ALSEP), which transmitted data back to Earth after the landing party had gone. They also collected more rocks.

Conrad and Bean walked over to Surveyor 3 and removed pieces, including the TV camera which they took back to Earth. When scientists took the camera apart in a clean room, they found a small colony of common bacteria *Streptococcus Mitis* inside the device. It was thought that the microorganism had somehow survived the vacuum of space, three years of exposure to radiation, and the freezing temperatures of the lunar night.

This was later discounted and it is now thought that the camera had been contaminated by the lax clean-room procedures on its return to Earth. Archive footage of the examination of the camera shows two investigators wearing masks and gloves. However, much of the head and neck were exposed and they were wearing short-sleeve scrubs that left the arms uncovered.

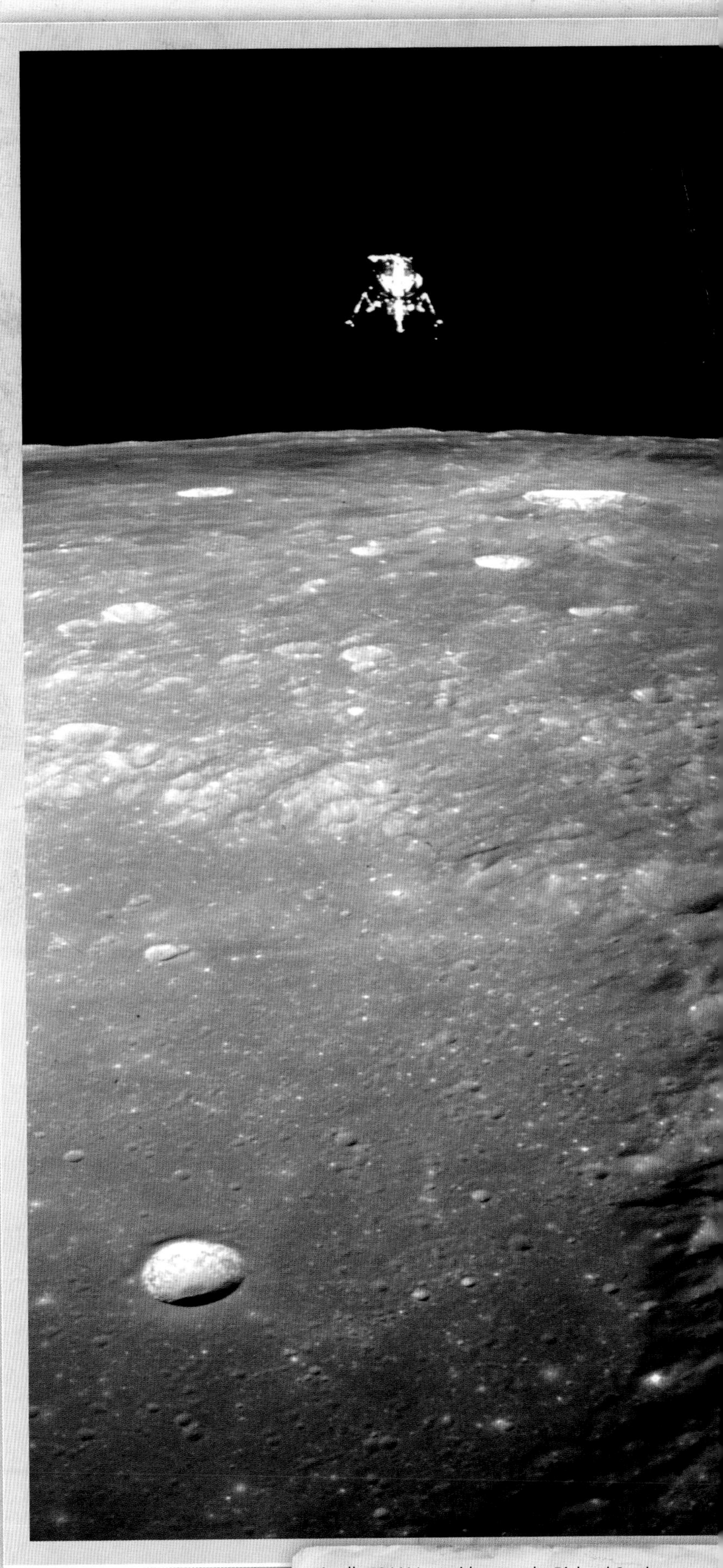

Apollo 12 LM *Intrepid* as seen by Richard Gordon in CSM *Yankee Clipper* shortly after separation. The LM with Charles Conrad and Alan Bean onboard is 68 miles (110 km) above the surface highlands heading into the Ocean of Storms.

The eclipse of the sun as seen from the Apollo 12 spacecraft during its trans-Earth journey home from the Moon.

## SOLAR ECLIPSE

After Conrad and Bean rejoined Gordon in orbit, the LM was left to crash land on the Moon, with the seismometers the astronauts had left behind registering vibrations for more than an hour. They stayed in orbit for an extra day taking photographs then, on their flight back to Earth, they witnessed a unqiue solar eclipse with the Earth traveling in front of the Sun.

After Apollo 12 entered the atmosphere, the explosive bolts did work and the parachutes deployed. There was one further accident. Bean was knocked unconscious during splashdown when a 16mm camera was dislodged from storage and hit him on the head. The Apollo 12 CM *Yankee Clipper* is on display at the Virginia Air and Space Center in Hampton, Virginia, while the camera from Surveyor 3 is at the National Air and Space Museum in Washington, DC. In 2002, astronomers thought they might have discovered another small moon orbiting the Earth. They designated it J002E3, but it turned out to be the third stage of the Apollo 12 Saturn V rocket.

## Surveyor 3

Launched on April 17, 1967, Surveyor 3 was the third unmanned lander NASA sent to the Moon. It landed in a large crater in the Oceanus Procellarum, or Ocean of Storms. Ranger 7, the first US space probe to transmit close-up images of the lunar surface to Earth, crash-landed there on July 31, 1964, giving the crater its name Mare Cognitum—"The Sea That Has Become Known." The Soviet Union's Luna 5 crashed in the area on May 12, 1965, after its retro-rockets failed.

Surveyor 3 was also supposed to make a soft landing, but the craft's radar was confused by highly reflective rocks on the Moon, causing it to bounce on the lunar surface twice, once to about 35 feet (11 m) and again to about 11 feet (3 m). On the third impact with the surface Surveyor 3 settled down to a soft landing as intended. Nevertheless, it continued to function.

Surveyor 3 carried a scoop on an extendable arm. This took soil samples that could be placed in front of the television camera it carried. The camera also transmitted back panoramic pictures of the lunar surface and shots of the spacecraft itself.

Surveyor 3 shut down for the lunar night on May 3, 1967, as its solar panel no longer produced electricity. It could not be reactivated after the next lunar dawn, having been damaged by the temperature of -391°F (-235°C) it had experienced.

The unmanned Surveyor 3 spacecraft as photographed during the Apollo 12 mission.

# THE MOON AND SCIENCE

The Moon is Earth's only natural satellite, and is the easiest celestial object to find in the night sky. It is roughly a quarter of Earth's size and consequently has a great effect on the planet and very possibly makes life on Earth viable.

The rhythm of the Moon's phases has guided humanity for millennia. As the Moon orbits around the Earth, the half of the Moon that faces the Sun will be lit up. The different shapes of the lit portion of the Moon that can be seen from Earth are known as phases of the Moon. Each phase repeats itself every 29.5 days roughly equal to a calendar month.

The Apollo ALSEP program answered many questions about the Moon including:

- How old is the Moon? *The moon is very old—a new analysis of lunar rocks brought to Earth by Apollo astronauts suggests that the moon formed 4.51 billion years ago, just 60 million years after the solar system itself took shape. It is believed that a giant impact from an asteroid about the size of Mars crashed into the primitive molten Earth, and knocked off the raw ingredients for the Moon into orbit.*
- What is inside the Moon? *NASA's seismographs left on the Moon by Apollo missions collected data that showed the Moon has an internal structure pretty much like the Earth, with a thin crust of about 40 miles (65 km), a mantle about 62 miles (100 km), and a core with a radius of about 310 miles (500 km).*
- What is the actual geometric shape of the Moon? *Officially the Moon is an oblate spheroid, otherwise known as a squashed sphere. The simplest evidence can be seen during solar eclipses when the Sun's shadow is always nearly circular. The only geometric object that can yield a near-circular eclipse in any orientation is a spheroid.*

# CHAPTER 19

# NASA'S FINEST HOUR

## APOLLO 13

After the first Moon landing, Apollo 12 seemed a bit of an anticlimax. Once NASA had shown that it could put a man on the Moon, doing it a second time made it appear almost routine. The worldwide audience was not interested in the scientific experiments that the astronauts laid out on the surface of the Moon or the questions they were designed to answer. However, with Apollo 13, the world was once more riveted to their televisions watching the space program.

On April 11, 1970, Apollo 13 lifted off for the Moon with Commander Jim Lovell, Command Module Pilot Jack Swigert, and Lunar Module Pilot Fred Haise aboard. At age 42, Lovell was NASA's most experienced astronaut. He had been on two Gemini missions—including a fourteen-day orbital marathon—and Apollo 8 which was the first manned mission to circle the Moon. He had clocked up 572 hours of spaceflight.

Fred Haise who was 36 years old, had been in the backup crew for Apollo 8 and 11, while 38-year-old Swigert had been in the support crew for Apollo 7. He had been on the backup crew for Apollo 13 but two days before liftoff he had replaced Ken Mattingly who have been exposed to German measles.

### SPARKING A CRISIS

There was a minor problem with the launch. In the second stage, one of the engines shut down two minutes early. However, the other four second-stage engines and the third-stage engine burned a bit longer to compensate and it almost achieved the scheduled 115-mile (185 km) Earth orbit. Apollo 13 set off for the Moon two hours later. It separated from the third stage of the Saturn rocket, then performed the transposition and docked the CM *Odyssey* with the LM *Aquarius*.

At almost fifty-six hours into the flight and some 200,000 miles (322,000 km) from Earth, Swigert received orders from Mission Control to turn on the fans that stirred up the hydrogen and oxygen cryogenic tanks in the SM. This sparked a short-circuit between exposed wires in the oxygen tank, causing it to explode with a loud bang.

"Houston, we've had a problem here," said Swigert. This is usually misquoted as "Houston, we have a problem," which was the line delivered in the 1995 movie *Apollo 13*. When Mission Control asked them to say it again, Lovell repeated: "Houston, we've had a problem."

## THE SOURCE OF THE PROBLEM

The problem could be traced back to 1962 when North American Aviation, later part of Boeing, was awarded the contract to build the Command and Service Modules for the Apollo program. North American Aviation subcontracted the construction of the cryogenic tanks to Beech Aircraft in Boulder, Colorado. In turn they subcontracted the thermostatic safety switches for the tanks. These were to operate at 28 volts.

In 1965, NASA decided to upgrade the spacecraft's electrical systems to make them compatible with the 65 volts used by the ground systems at Cape Kennedy. However, the subcontractor was never informed of the change.

> *There was a point where panic almost overcame me.*
>
> Sy Liebergot, Flight Controller

The two oxygen tanks on Apollo 13 were scheduled to fly on Apollo 10, but were swapped. During the changeover, oxygen tank 2 was dropped from a height of about two inches. It was inspected and appeared to be intact before it was installed on Apollo 13. During a test it was filled with liquid oxygen, but failed to discharge when gas was pumped in. So it was decided to heat the tank and boil off the oxygen. But when 65 volts was applied to the heaters, the switches were fused shut. Instead of tripping when the temperature reached 80°F (27°C), the heaters kept running and the temperature climbed to 930°F (500°C). This burned the insulation off the wires around the heaters.

When the fans were turned on with Apollo 13 in space, the bare wires sparked and the remaining insulation burned in the pure oxygen. The temperature and pressure soared. The explosion ruptured the tank and blew off some of the outer skin of the SM—though neither the astronauts in the CM nor the engineers at Mission Control could see this.

The shortage of oxygen starved the fuel cells, which stopped producing electricity. Before the backup batteries cut in, the onboard computer reset and the high-gain directional antenna cut out. The data received back in Houston was garbled and Mission Control assumed that there was an instrumentation problem.

Both Mission Control and the astronauts tried to identify and rectify the problem, but soon it became clear that the problem was more serious than they first thought.

"It looks to me, looking out of the hatch, that we were venting something out into space," Lovell reported.

Liquid oxygen was squirting out into space.

"It was not an instrumentation problem but some kind of monster systems failure," said flight controller Sy Liebergot, in charge of the electrical systems. "It was probably the most stressful time in my life."

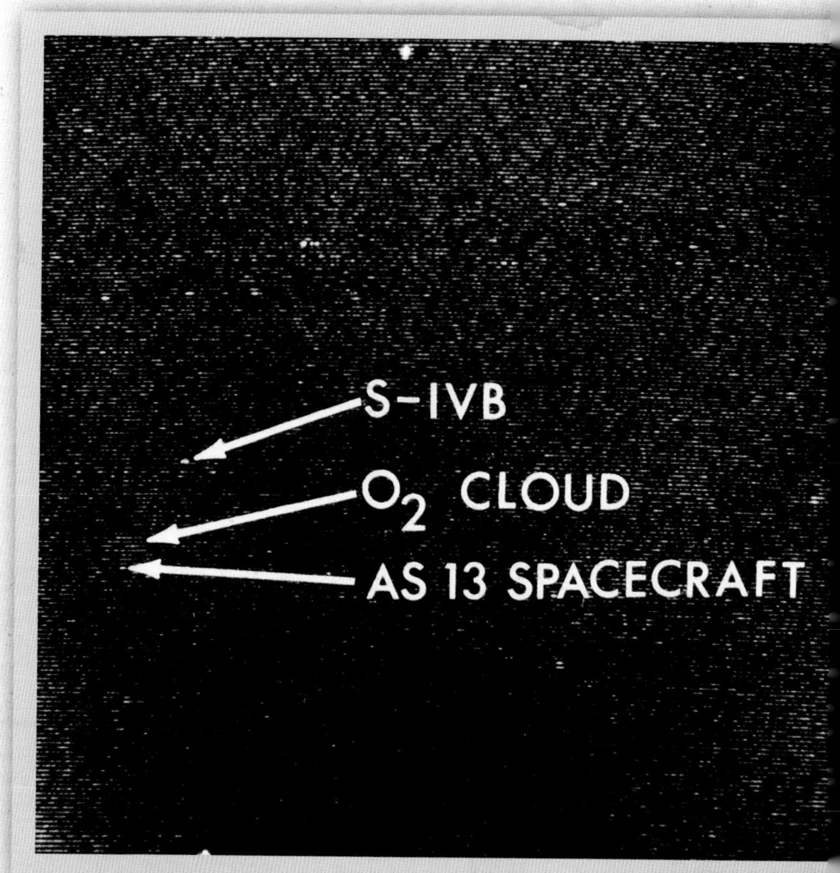


A telescopic photograph showing the Apollo 13 spacecraft in trans-lunar trajectory in the distant sky.

## KEEPING COOL UNDER PRESSURE

"Okay, let's everybody keep cool," said Flight Director Gene Kranz. "Let's solve the problem, but let's not make it any worse by guessing."

Liebergot suggested that they isolate the surge tank. This was a small reserve tank of oxygen that the crew breathed during reentry after the SM had been jettisoned. There was a danger that, with other reserves of oxygen being lost, this would be drained too. They also powered down everything in the CM, apart from the guidance system.

The last fuel cell on the *Odyssey* was running low on oxygen and the race was on to switch on the systems in the LM which they intended to use as a "lifeboat." During the journey to the Moon, the LM was powered via umbilicals from the CSM which was now running out of power and power was needed to operate the relays that connected the large batteries in the descent stage of *Aquarius*. Instead power would have to be tapped from the smaller batteries in the ascent stage. This had never been done before, nor had anyone planned for it. A backroom team at Mission Control began to work out a set of step-by-step procedures.

Working from the wiring diagrams of the LM, they came up with a list of some fifteen switch throws. Once the power source was switched from the now-dead umbilicals to *Aquarius's* batteries they could then power up the LM in lifeboat mode. Some thought had already been put into this. On Apollo 10, a simulation had been run on what could happen if the fuel cells failed in the same place that the tank exploded on Apollo 13. Unfortunately, they were unable to get the LM powered up in time and the simulation ended with the crew dead.

> *We needed to buy some time so that when we did make a move, it would be the proper move.*
>
> Gene Kranz, Flight Director

The nub of the problem was to work out how to reverse the power so it flowed from the LM into the CSM, rather than the other way around. More thought had been put into the problem during Apollo 11 and Apollo 12. The crew of Apollo 13 had a copy of the official LM activation checklist onboard. Flight controllers quickly cut this down so that the three astronauts could get onboard *Aquarius* and power her up while there was fifteen minutes of life left in the last fuel cell on *Odyssey*.

Flight controllers gather around the console in the Mission Operations Control Room (MOCR) during the problem-plagued Apollo 13 mission.

# GENE KRANZ

Born in Toledo, Ohio, in 1933, Eugene Francis "Gene" Kranz wrote a high-school thesis titled: "The Design and Possibilities of the Interplanetary Rocket." In 1951, he went on to study Aeronautical Engineering at Saint Louis University's Parks College of Engineering, Aviation and Technology. Graduating with a Bachelor of Science degree in 1954, he was commissioned as a second lieutenant in the US Air Force Reserve, completing his pilot training at Lackland Air Force Base in Texas.

After a tour flying F-86 Sabres in Korea, he quit the Air Force and went to work for McDonnell Aircraft Corporation, researching and testing new surface-to-air and air-to-ground missiles at the USAF Research Center at Holloman AFB.

In 1960, he answered a "help wanted" ad in *Aviation Week* and went to work for NASA's newly formed Space Task Group at the Langley Research Center in Hampton, Virginia. For the next four years, he worked in the Flight Control Operations Branch developing and writing rules used by flight directors during manned spaceflight missions.

When the Manned Spacecraft Center opened in Houston, Kranz moved to Texas and became chief of the Flight Control Operations Branch. He served as Gemini flight director from 1964 to 1968. Then he became flight director for Apollo, working in that role for Apollo 11's lunar landing and the return of the crippled Apollo 13. From 1969 to 1974, he also served as the flight operations director during the Skylab program. At the end of the Skylab program, he was promoted to deputy director of Flight Operations and then in 1983 to director of Mission Operations, retiring in 1994.

His 2000 autobiography was called *Failure Is Not An Option*. Actor Ed Harris, playing Kranz in the 1995 movie *Apollo 13*, uttered the phrase. It was the invention of screenwriter Bill Broyles and not something that Kranz had ever said.

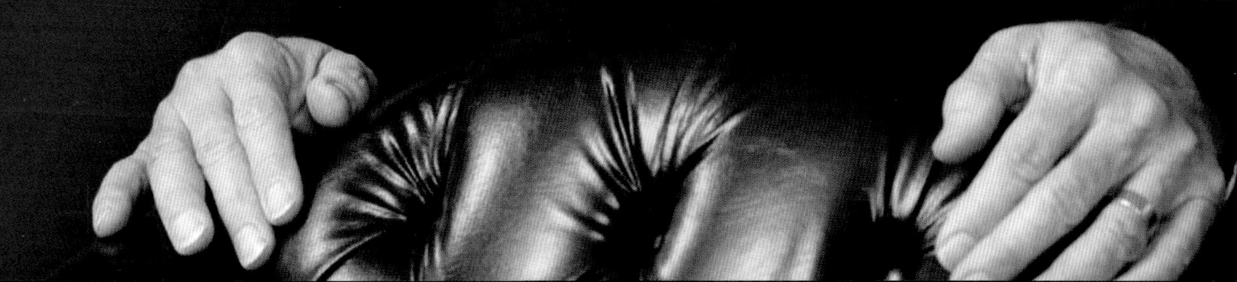

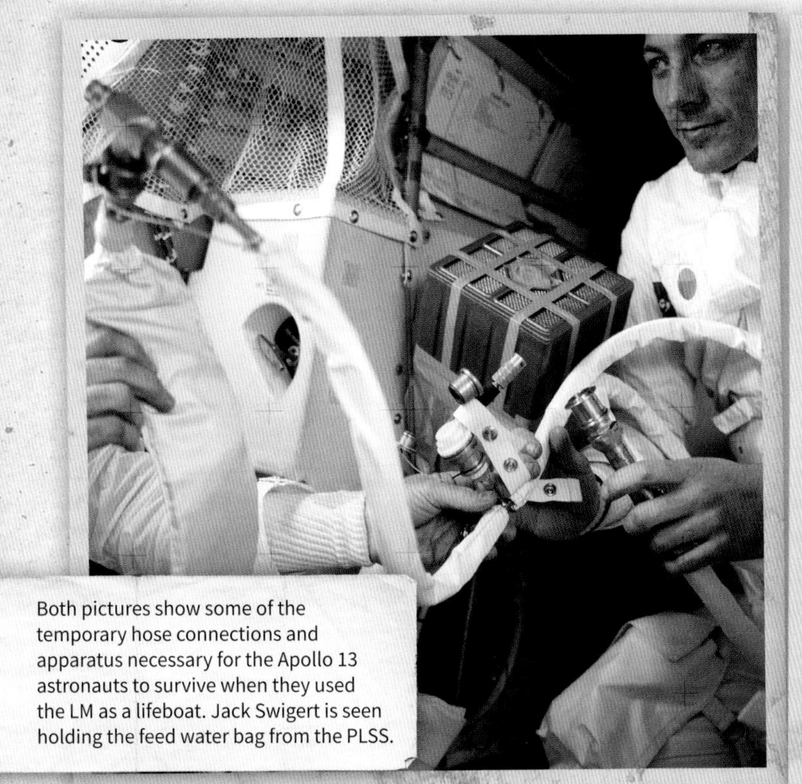

Both pictures show some of the temporary hose connections and apparatus necessary for the Apollo 13 astronauts to survive when they used the LM as a lifeboat. Jack Swigert is seen holding the feed water bag from the PLSS.

## GET THE ASTRONAUTS HOME

Then Mission Control had to decide on how to get the astronauts home. The standard abort procedure was to fire the main engine on the SM and send the spacecraft directly back to Earth. This would use every last drop of fuel and had to be executed perfectly. Also it was not known what damage had been done to the main engine by the explosion in the SM. Firing it might set off a catastrophic explosion and blow the whole spacecraft up.

The other option was to let Apollo 13 travel on to the Moon, slingshot around it, and come back to Earth that way. That would take four days with three men in the LM, which had only been designed to support two men for two days, but Kranz decided this was the better option as it bought them time.

"I was pretty much betting that this control team could pull me out of the woods once we decided to go to the Moon," he said.

The next problem was that Apollo 13 was on a "no-free-return trajectory." Once it had slingshot around the Moon and headed back, it would miss the Earth by several thousand miles. To get the spacecraft on a free-return trajectory to make a splashdown, an engine had to be fired. The only one available was the one on the LM's descent stage. No one had envisaged doing this under such circumstances, so fresh software had to be written. Nevertheless in a couple of hours the IBM mainframes at the Manned Spacecraft Center's Real Time Computer Complex came up with a free-return maneuver.

## OLD-FASHIONED ARITHMETIC

To conserve enough power to make it back to Earth everything in the CM was going to have to be switched off. That included the guidance system. But it was needed to make the free-return maneuver. Instead the guidance system on the LM would have to be used, but as it was switched off during translunar flight, it had no idea which way the spacecraft was pointing. The crew would have to transfer the alignment data from the guidance system onboard *Odyssey* to the system on *Aquarius*.

To do this, they had to read the angles out of one computer and type them into the other. However, *Odyssey* and *Aquarius* were docked head to head, so the angles had to be inverted. This had to be done by old-fashioned arithmetic. To make sure the astronauts got the figures right, the math was checked by Mission Control using paper and pencil. Once the data had been transferred, all the systems in the CM were switched off.

Later, when Apollo 13 was approaching Earth, everything would have to be switched back on again. The usual power-up procedures for the CM started two days before the return journey. This time they would be short of power, so the power-up would have to be done in a couple of hours. A team got down to shortening the procedures. They came up with a checklist five pages long.

## FREE-RETURN TRAJECTORY

The burn of the LM's engine went off without a hitch, putting the spacecraft on a free-return trajectory. However, it would have to splashdown in the Indian Ocean where there was no recovery team to pick up the CM. The flight dynamics controller came up with a plan to shave twelve hours off the flight, so they would splashdown in the Pacific.

Another team worked on saving energy in the LM. This included switching off the heating which caused a dramatic drop in temperature. The crew would be freezing, but the spacecraft's power consumption would go down to around 12 amps—about as much as that used by a household vacuum cleaner.

The LM's descent engine had to be fired again two hours after Apollo 13's closest approach to the Moon. Once this was done, the LM's guidance system would also have to be switched off and the crew would essentially be flying blind.

Although there was enough oxygen onboard for the crew to breath, carbon dioxide levels were rising. Lithium hydroxide crystals were used to absorb the gas from the air. However, the canisters onboard *Aquarius* were round and the spare ones on *Odyssey* were square. Engineer Ed Smylie had already realized that this was going to be a problem. He worked out how to connect them using things he knew were onboard. These included duct tape, the cover of a flight manual, a plastic bag, and a sock. Miraculously, the makeshift adapter worked.

## DRIFTING OFF COURSE

The second burn had been perfect, but the spacecraft was drifting off course. It was later determined that a water-cooling vent on the LM was acting like a tiny jet. A corrective burn was required, but they had no guidance system. However, one of the flight controllers recalled that on Mercury, Gemini, and Earth-orbiting Apollo missions, astronauts had used the terminator—the line when day turns into night—as a reference point and they were able to pull off the corrective maneuver. It had to be unerringly accurate. If they missed the entry corridor by as much as a degree, they risked burning up in the atmosphere or bouncing back into space.

Later they had to make another course correction at the behest of the Atomic Energy Commission (AEC). The LM carried plutonium to fuel the ALSEP. It was an isotope that created chain reactions, like those used in a bomb, and generated energy from the heat of radioactive decay. It was in a protective case that would prevent it burning up in the Earth's atmosphere, but the AEC still insisted the LM had to splashdown as far away from anywhere as possible. The course adjustment would direct the remains of the *Aquarius* to the bottom of the Tonga Trench, 6.5 miles (10.6 km) down in the Pacific.

The severely damaged Apollo 13 Service Module (SM) was photographed from the LM/CM after the SM was jettisoned.

## THE GAMBLE PAYS OFF

Having shortened the flight by twelve hours, it became clear that Apollo 13 was going to make it back to Earth. Engineers had now worked out how to restore the power again. But no one knew how much electricity it would take to power up the CM. There was a risk that the batteries would run out before splashdown.

Normally, the instrumentation would be turned on before power-up, so that each step could be checked to see if it was successful. To save power, switching on the instrumentation would be delayed until after all the other systems had been turned on. In that case, if any errors had been made it would be too late to fix them. But the three astronauts had been so well-trained that the gamble paid off.

> *Farewell,* Aquarius, *and we thank you.*
>
> James Lovell

## BIDDING A FOND FAREWELL

Preparing for reentry, the SM was jettisoned. As it tumbled away, the crew could at last see the damage that the exploded oxygen tank had caused. One side of the module was missing and it looked as if the engine had been damaged too. Kranz had been right. Had they followed the standard abort procedure, the SM and the rest of Apollo 13 would have blown up.

The Apollo 13 spacecraft heads toward a splashdown in the South Pacific Ocean.

Finally, with the astronauts back in the CM, *Aquarius* had to be jettisoned too. They bid her a fond farewell. She had saved their lives.

Two hours later, *Odyssey* hit the outer reaches of the atmosphere and the astronauts and Mission Control prepared for the customary communications blackout. Neither had any idea whether the heat shield had been damaged in the explosion. Before Apollo 13 plunged into the sea of static, Swigert radioed: "I know all of us here want to thank all of you guys down there for the very fine job you did."

Kranz recalled: "It was the worst time of the whole mission ... You asked yourself 'did I give the crew everything I needed to and was the data right?'"

Then the plane circling the splashdown area picked up a signal. But still the parachutes had not deployed. There was a danger that they had frozen solid when the heating system was turned off. Then the *Odyssey* finally appeared on the large screen in Mission Control which was taking live TV coverage from USS *Iwo Jima*. Soon Lovell, Swigert, and Haise were seen walking on its deck. Against all the odds, NASA had achieved the impossible.

## RECORD BREAKING MISSION

Despite great hardship caused by limited power, loss of cabin heat, shortage of potable water, and the critical need for makeshift repairs to the carbon dioxide removal system, they had got the astronauts back, alive, from the stricken Apollo 13, six days after launch.

Apollo 13, following the free-return trajectory, reached an altitude of 137 miles (254 km) over the far side of the Moon, approximately 60 miles (100 km) greater than the orbital altitude on all the other Apollo lunar missions. At the same time, the Moon was almost at apogee which increased the distance the spacecraft had to travel from Earth. The combination of the two effects ensured that Apollo 13 reached a distance from home of 248,655 miles (400,171 km), a spaceflight altitude record marking the farthest humans have ever traveled from Earth.

Jim Lovell called the mission a "successful failure," due to the safe return of the astronauts, but with the aborted lunar landing. However, with several dramatized versions of the mission including the movie *Apollo 13* (1995), it has also been remembered as "NASA's finest hour."

CHAPTER 20

# MISSION TO FRA MAURO

## APOLLO 14

Apollo 14 took over the mission that Apollo 13 failed to fulfill, which was to explore the Fra Mauro region of the Moon. Indeed the crew—Alan Shepard, Stuart Roosa, and Edgar Mitchell—had originally been slated for Apollo 13. But Shepard, who had been the first American in space, had then been diagnosed with Ménière's disease, a condition in which fluid pressure builds up in the inner ear. He was grounded and made Chief of the Astronaut Office responsible for astronaut training.

In 1969, he underwent risky surgery to relieve the condition and was restored to full flight status. He was put down to command the next Moon mission and was teamed with two new recruits—Command Module Pilot Stuart Roosa and Lunar Module Pilot Edgar Mitchell, who had both served on the support crew on Apollo 9. With less than 15 minutes 30 seconds of spaceflight time between them, they were known as "the three rookies" and were pushed back from Apollo 13 to Apollo 14 so they could receive extra training.

The crew of Apollo 14. Left to right: Edgar D. Mitchell, Alan B. Shepard Jr., and Stuart A. Roosa.

# SKYLAB

America's first space station, Skylab was launched into low-Earth orbit on May 14, 1973, by an unmanned Saturn V whose third stage had been adapted to house a workshop and living space. Around a minute into the launch a thermal-meteoroid shield was ripped off, taking one of the solar power arrays with it and the other was jammed in a folded position. Without the shield Skylab risked overheating and the failure of the solar panels left it short of power.

The manned missions to the space station were to use an Apollo CSM blasted into space on top of a smaller Saturn 1B rocket. The first, which was scheduled for the next day, was postponed until May 25 while NASA engineers figured out how to repair the damage. The first job of its three-man crew—Charles Conrad, Joseph Kerwin, and Paul Weitz—was to get it working again. First they deployed a solar shield through a small airlock for scientific experiments on the side facing the Sun. Once outside, the makeshift shield opened like an umbrella and the temperature in the space station dropped to bearable levels. It also generated much-needed electricity.

Next Conrad and Kerwin had to make a spacewalk to cut away some metal that was jamming the solar wing in its folded position. They stayed onboard for twenty-eight days, doubling the fourteen-day US record Borman and Lovell had set on Gemini 7 and proving that human beings could live and work in space for extended periods. During their time onboard, they studied processing materials in microgravity and made observations of the Earth and Sun.

A second crew—Alan Bean, Owen Garriott, and Jack Lousma—visited on July 28, 1973. Garriott and Lousma performed a spacewalk to erect a new twin-pole solar shield that provided better thermal control for the remainder of the Skylab missions. They remained onboard for fifty-nine days.

On the next mission blasting off on November 16, astronauts Gerald Carr, Edward Gibson, and William Pogue stayed on Skylab for eighty-four days. The three missions produced a vast study of the Earth—its crops, weather, and changing environment—and completed a detailed study of the Sun. The crews also manufactured alloys and grew perfect crystals.

There were plans to use Skylab again, but its orbit decayed faster than expected. On July 11, 1979, it entered the Earth's atmosphere and broke up, scattering debris across the south-eastern Indian Ocean and Western Australia.

# APOLLO-SOYUZ

In the spirit of John F. Kennedy, NASA had long sought cooperation in space with the Soviet Academy of Sciences. As American involvement in the Vietnam war came to a close and the Cold War began to thaw, President Nixon and Soviet Premier Alexei Kosygin signed an agreement permitting this to go ahead.

On July 15, 1975, Soyuz 19 was launched from the Baikonur Cosmodrome in Kazakhstan carrying cosmonauts Alexei Leonov and Valery Kubasov. Hours later an Apollo CSM on top of a Saturn 1B rocket carrying Thomas Stafford, Vance Brand, and Donald Slayton blasted off from Cape Kennedy in Florida. Over the following two days, both made orbital adjustments bringing them into a circular 142-mile (229-km) orbit.

On July 17, the two craft docked high over the Atlantic Ocean. The hatches were opened and the two crews swapped greetings and gifts, and shared a meal. The following day, Brand joined Kubasov in the Soyuz, while Leonov joined Stafford and Slayton in the Apollo. After giving TV viewers a tour of each vehicle, the crew members conducted science experiments and ate together again. Later, Kubasov and Brand left the Soyuz to join Slayton in the Apollo, leaving room for Leonov and Stafford to spend time in the Soyuz.

After the crews retreated to their own craft, the hatches were closed and Apollo and Soyuz undocked. The Soviet craft landed back in Kazakhstan safely on July 21, while Apollo splashed down in the Pacific on July 24.

Apollo 14 was to be the last of the H missions—that is, lunar missions with a short stay on the surface and two EVAs. It lifted off on January 31, 1971, after a three-month delay so that all the recommendations made by the Apollo 13 Review Board could be instituted. All combustible materials were removed from the oxygen tanks and changes were made to the fan system. A third oxygen tank was added, away from the other two. It had extra isolation valves to prevent it leaking if there was another explosion.

Extra batteries were installed in the CM and the power transfer system between the CM and the LM was changed to make the reversal of the flow easier. Apollo 14 also carried extra water as it had run low on Apollo 13.

Apollo 14 was seen as the beginning of the end for the Apollo program. Apollo 20 had been canceled in January 1970 and Apollos 18 and 19 that September. Some of the equipment would be used in Skylab—the US space station that orbited the Earth from 1973 to 1979—and the joint US-Soviet Apollo-Soyuz Test Project, where an Apollo CSM docked with a Soyuz 19 in July 1975.

The docking port of the LM *Antares* which caused CSM *Kitty Hawk* so much difficulty during Apollo 14 transposition and docking.

## DOCKING PROBLEMS

The launch of Apollo 14 was delayed by forty minutes due to the weather. When it eventually happened, the launch and the translunar injection went perfectly. But when the CSM separated from the third stage of the Saturn V and tried to extract the LM, there was a problem. Roosa turned the CSM around, aligned the docking probe, and moved slowly toward the docking port. However when they met, the initial soft docking did not occur. Instead the CSM bounced away.

Roosa was perplexed. This was a maneuver he had practiced endlessly. Mission Control assured him there was no problem and told him to try again. He did, but the two spacecraft failed to dock again. It was difficult to tell if anything was wrong. Checking the docking probe was a difficult procedure. The cabin would have to be depressurized while the CM *Kitty Hawk*'s probe was removed and inspected. If the CSM was brought close enough to the LM *Antares*, it should be possible to inspect the docking port too, but this was an untried procedure that Mission Control were unlikely to approve. If they could not dock, the mission would be over.

Instead Mission Control told Roosa to try docking again, but this time to keep firing the thrusters. While the CSM pushed against the LM, he should then retract the docking probe and see if the twelve hard-docking latches engaged. He did this and it worked, much to the relief of the crew. Once they were safely docked, the probe and port were dismantled, but there was no indication why the docking had not worked the first time.

## MAKING THE DESCENT

There was another problem after *Antares* and *Kitty Hawk* separated. The circuit breakers on the antenna popped a couple of times giving them communications problems. Then the "abort" light came on in the LM. Mission Control told Mitchell to hit the instrument panel to see whether it would go off. It did go off, but then flashed on again.

It was assumed that the problem was caused by loose pieces of solder making random contact with the circuit board. Fortunately this had happened before the descent was made, otherwise the computer would have automatically aborted the landing. The equipment had been made by the Massachusetts Institute of Technology and an MIT expert was on hand. He came up with a way to modify the program so that it did not abort the landing on a spurious "abort" signal. Mitchell had to key these modifications in by hand.

Then, as they descended toward the surface, the landing radar did not come on. Mission Control told them to reset the circuit breaker. Shepard had already decided to make the landing even without the radar, though this was forbidden. Fortunately, the landing radar did blink on when they were just 15,000 feet (4,600 m) from the surface. Despite all the problems, they made a good landing.

Alan Shepard left the LM first. He was the only Mercury astronaut to make it to the Moon and, at age 47, the oldest person to do so. He and Mitchell took samples and deployed the ALSEP. This was done with a new piece of equipment called a Modular Equipment Transporter (MET). Then they set up the flag.

## SEARCHING FOR CONE CRATER

After a meal and a fitful sleep, Shepard and Mitchell set out to Cone Crater. They had photographs taken by previous Apollo Command Module pilots to guide them, but in the uniformly gray vista it was difficult to make out any landmarks, some of which were hidden in craters.

They took the MET with them, loaded with the tools they would need to take samples. Although it saved them carrying the tools and samples, they found the MET unwieldy as the surface was strewn with rocks. Every time the wheels of the MET hit one, in the low gravity, it flew in the air.

Several times they thought they had reached the rim of Cone Crater, only to find they were mistaken. Mission Control gave them more time, but eventually they ran short of oxygen. The trek was curtailed, though they did manage to collect nearly 100 pounds (45 kg) of rock and dust. Later analysis showed that they had come within 100 feet (30 m) of the crater and some of the samples they collected were ejecta from it. In all they had traveled almost nine-tenths of a mile (1,400 m) and, for the first time, astronauts on the lunar surface had been out of sight of the LM. They returned, cursing in frustration.

## Modular Equipment Transporter

The Modular Equipment Transporter (MET) was a two-wheeled vehicle that was used to transport equipment from the LM and to carry lunar rock samples back to it. Nicknamed the rickshaw, it carried the ALSEP and photographic equipment. However, it proved difficult to use in rough terrain and in Apollo 15 it was superseded by the motorized Lunar Roving Vehicle.

Alan B. Shepard Jr. in lunar surface training at the Kennedy Space Center with the modular equipment transporter (MET).

## LUNAR GOLF AND ESP

After taking more samples and checking the ALSEP, the mood lightened. Shepard pointed a TV camera at himself and said: "You might recognize what I have in my hand as the handle for the contingency sample return. It just so happens to have a genuine six iron on the bottom of it. In my left hand, I have a little white pellet that's familiar to millions of Americans."

He was going to play golf. Unfortunately the EVA suit was so stiff he could not make a two-handed shot, but he made a number of one-handed shots. The final shot he said went "miles and miles." In fact the ball only traveled about 200 yards (183 m).

The liftoff of the *Antares* from the lunar surface went smoothly and they had no difficulty docking back with the *Kitty Hawk*. They transferred the samples to the CSM and stowed the docking probe, which would normally have jettisoned with the LM, so they could take it back to Earth to see why it had given them problems with docking on the way to the Moon.

On the way back to Earth, Mitchell tried some experiments with extra-sensory perception. He stared at a series of shapes for fifteen seconds each in an attempt to transmit the image back to would-be psychics on Earth. The results were open to interpretation, but were generally thought to be a failure. Disgraced medium Olof Jonsson even claimed to have received a message from Mitchell on a day he had not transmitted one.

After splashdown the astronauts were put in a mobile quarantine facility on USS *New Orleans*. They were the last crew to undergo this as the risk of contamination from outer space was then dismissed.

*Unfortunately the suit is so stiff, I can't do that with two hands, but I'm going to try a little sand trap shot here.*

Alan Shepard, Apollo 14
Golfing on the Moon, February 6, 1971

Alan Shepard pulled out a makeshift six-iron he smuggled onboard Apollo 14 and hit two golf balls on the lunar surface, becoming the first—and only—person to play golf anywhere other than Earth.

# CHAPTER 21

# MOUNTAINS OF THE MOON

## APOLLO 15

Apollo 15, which explored the Moon in July and August 1971, was the fourth manned landing and was the first Apollo mission to carry a Lunar Rover. However, the Soviet Union had beaten America to the punch again. The first successful rover to land on the Moon was in fact the Soviet Union's robotic Lunokhod 1. It rolled down a ramp from the Luna 17 lander on the Sea of Rains on November 17, 1970.

During the following ten months the eight-wheeled vehicle traveled over 6 miles (10 km) on the lunar surface, transmitting geological data and imagery back to Earth. Solar-powered during the day, a radioisotope heater helped Lunokhod 1 endure the freezing-cold temperatures of the Moon at night. However, like all Soviet missions to the Moon, it was unmanned and operated by remote control from Earth. NASA's "Moon buggy" carried the two-man crew of the LM and was driven hands-on by the astronauts.

### LUNAR GEOLOGY TUTORIAL

The crew of Apollo 15 was to be Commander David R. Scott, Command Module Pilot Alfred M. Worden, and Lunar Module Pilot James B. Irwin. Harrison "Jack" Schmitt, a qualified geologist selected as an astronaut in 1965, brought in Lee Silver, his old teacher at Caltech, to train Scott and Irwin to be proficient field geologists. He had already trained Lovell and Haise from Apollo 13, though they did not have the chance to put their learning into use.

Schmitt and Gordon from the backup crew went on sixteen field trips with Scott and Irwin, learning to spot what may be significant rocks. After the mission, Don Williams, a member of the science team said of David Scott: "Geologists who worked with him are unstinting in their praise of his interest and ability in their subject [that] blossomed into excitement and total commitment."

Apollo 15 carried two new things. A Lunar Rover was stowed in the equipment bay of the descent stage of the LM, while the spare compartment of the SM carried the new Scientific Instrument Module (SIM), which allowed Worden to make a detailed survey of the Moon. One camera was powerful enough to make out the LM and the Lunar Rover on the surface 60 miles below (100 km). Worden got a geology tutor of his own, the Egyptian Farouk El-Baz, known affectionately as "The King."

## HADLEY-APENNINE

With only three more Apollo missions scheduled, it was important to pick the right landing site. Hadley-Apennine, a region on the near side of the Moon along the eastern edge of the Mare Imbrium or Sea of Rains, was chosen. Some four billion years ago a massive asteroid had struck the Moon, creating the Imbrium Basin and piling up the Apennine range which reached 15,000 feet (4,500 m). Lava flows then carved the Hadley Rille, a channel which is 1,000 feet deep (300 m) and three-quarters of a mile (1.2 km) wide. Geologically this made it a fascinating region, though the approach would be over dangerously steep mountains.

At 9:34 a.m. on July 26, 1971, Apollo 15 blasted off from Launchpad 39A at Cape Kennedy after an uneventful countdown. It was estimated that a million spectators turned out, though the temperature that day was 85°F (30°C).

The mission carried a new enhanced LM, called *Falcon* after the official mascot of the Air Force Academy. The CSM was called *Endeavour*, after the ship Captain James Cook sailed in to make his three voyages of exploration in the Pacific Ocean. Carrying the LRV and the SIM, Apollo 15 was the heaviest manmade object that had ever orbited the Earth.

> *It was one of the happiest moments of my life ... at last it was my turn, at last I was leaving Earth and going to the heavens.*
>
> James B. Irwin, Apollo 15
> Lunar Module Pilot

## The SIM Bay

Apollo 15 was the first mission to carry a Scientific Instrument Module (SIM). It contained a panoramic camera, a modified version of the KA-80A camera carried by the US Air Force's spy satellites, along with a mapping camera, a laser altimeter, a gamma ray spectrometer, and a mass spectrometer. Apollo 16 carried the same configuration. These SIM bays also released small sub-satellites into lunar orbit to study the plasma, particle, and magnetic field environment of the Moon and map the lunar gravity field.

On Apollo 17, the particle and field sub-satellite was replaced with a Lunar Sounder Experiment, also known as the Coherent Synthetic Aperture Radar (CSAR). This used radar to study the structure of the top 1.25 miles (2 km) of the Moon's crust. The results were recorded on an optical recorder which took the place of the mass spectrometer. The gamma ray spectrometer was replaced by an ultraviolet spectrometer and the alpha and X-ray spectrometer by an infrared spectrometer.

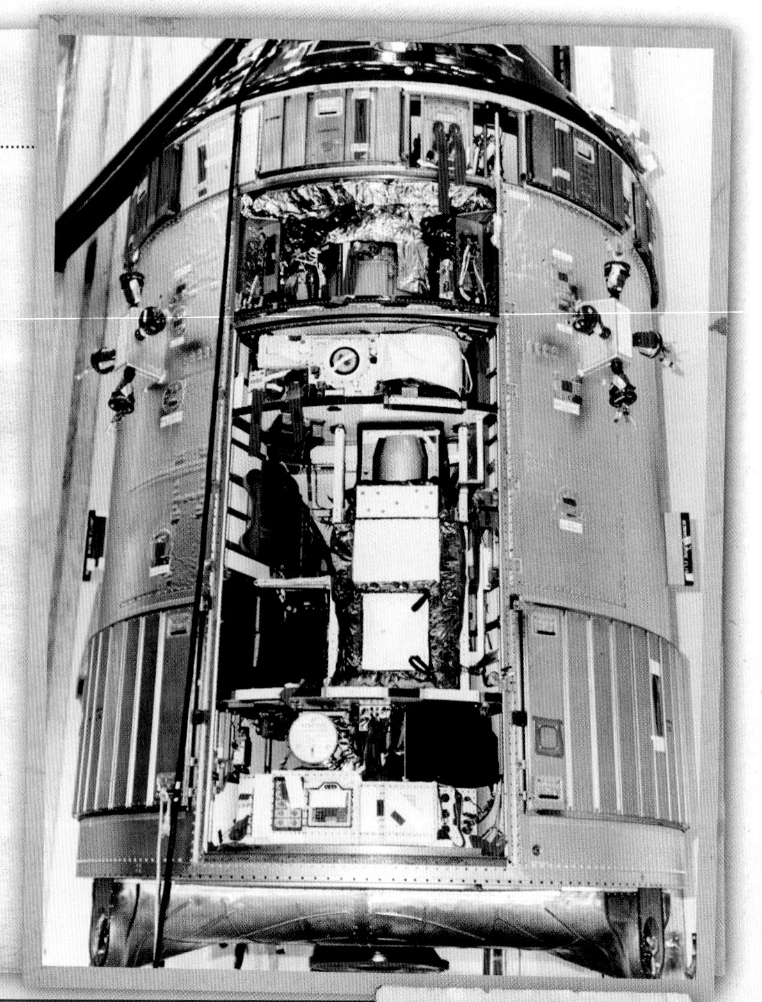

A close-up view of the Scientific Instrument Module (SIM).

## THE VIOLENT LAUNCH

David Scott, who had ridden a Saturn V before, remembered accelerating from zero to 18,000 mph (29,000 kph) in less time than it takes to drink a cup of coffee:

> *I would suspect from what I heard from the other guys that every launch was different. The configurations are only slightly different, but you have different times of day, different wind conditions, who knows? But from what I can recall the Apollo 9 launch was pretty violent, especially at staging. Fifteen wasn't quite that violent, but it was pretty violent.*

Worden and Irwin were taken unawares by the sudden jolt on separation. Scott grinned and said: "Sorry, I forgot to tell you about that."

Three hours after the launch, the S-IVB engine sent them on their way to the Moon. Then came separation and transposition. But soon after *Falcon* had been removed from its housing, a warning light lit up indicating that there was a serious problem with the mothership's main SPS engine. If this was not just another instrumentation malfunction, the landing would have to be aborted and the LM would have to stay attached to provide a backup to get the astronauts back to Earth.

Between them, the crew and Mission Control discovered a short circuit in the switch. They isolated it and Scott made the burn needed for the midcourse correction manually. This worked perfectly. Next they had a leak of drinking water, but Scott managed to tighten the loose joint.

## TOWERING PRECIPICES

They entered lunar orbit seventy-nine hours into the flight. After another four hours they made another burn and dropped into an elliptical orbit with a low point of 56,000 feet (17,068 m). Then they had a sleep, with Mission Control predicting that the orbit would further decay until the low point was 50,000 feet (15,240 m), but when they woke the orbit had a low point of 33,000 feet (10,058 m), plus or minus 9,000 feet (2,743 m).

This was a little too close for comfort as they were coming in over the Hadley-Apennines, so the *Endeavour* fired its thrusters to push it up to 50,000 feet (15,240 m), the planned height for *Falcon* to begin its descent. To clear the mountains, the LM had to come in at an angle of 26°, nearly twice as steep as the previous Moon landings.

As the *Falcon* descended, the radar told them that they were too high and Scott feared they were going to overshoot the rille. At 6,000 feet (1,828 m) the LM pitched forward 30° and suddenly mountainous precipices were towering 7,000 feet (2,133 m) above them.

This was a surprise as they had seen nothing like this in the simulations. Then they saw a canyon and recognized it as Hadley Rille. Still the landing site looked nothing like the simulations. Apparently, this was because the photographs used were not high resolution, so the detail had been lost.

Nevertheless they had to get down on the surface before they ran out of propellant. Scott was piloting the LM. He spotted somewhere relatively flat, but by the time the *Falcon* got within 60 feet (18 m) of the surface the engines stirred up so much dust that he could not see anything. All he had to guide him was Irwin reading out the height and rate of descent.

Suddenly a blue light flashed on the instrument panel. Irwin said "contact" and Scott cut the engines. The LM dropped the last few feet and landed with a thump. They heard a round of applause from Mission Control. Scott and Irwin were so well trained that they made the landing look easy. For the worldwide TV audience, there was little excitement. It seemed routine.

## WONDERS OF THE UNKNOWN

The LM had come down on the edge of a small crater, leaning over at 11° with the bell of the engine resting on the rim, but Scott was unfazed.

"Hadley Base here," he said. "Tell those geologists in the backroom to get ready because we've really got something for them."

Again the astronauts were supposed to sleep before they made an EVA, but just two hours after landing they removed the docking hatch above their heads and Scott put his head out. While taking some photographs, he gave a description of what he saw for those listening to *Voice of America*:

> *All of the features around here are very smooth. The tops of the mountains are rounded off. There are no sharp jagged peaks or no large boulders apparent anywhere.*

# THE LUNAR ROVER

The Apollo Lunar Roving Vehicle (LRV) was an electric buggy which meant that the astronauts could extend the range of their EVAs. Built of the lightest materials, it had a mass of 463 pounds (210 kg), so weighed just 77 pounds (35 kg) on the Moon, and was strong enough to carry an additional payload of 1,080 pounds (490 kg), weighing 180 pounds (82 kg) on the Moon. Its frame of aluminum alloy tubes was 10 feet (3.1 m) long with a wheelbase of 7 feet 6 inches (2.3 m). It was hinged in the middle so that it could be folded and hung in the LM quad 1 bay. Four lunar rovers were built, one each for Apollos 15, 16, and 17 and one that was used for spare parts, at a final cost of $38 million—the equivalent of $ 236 million in 2018.

The wheels consisted of a spun aluminum hub and a tire 32 inches (81.8 cm) in diameter and 9 inches (23 cm) wide. This was made from zinc-coated woven steel strands 0.030 inches (0.083 cm) in diameter. Titanium chevrons covered half the contact area to provide traction. Both sets of wheels were steerable, giving a steering radius of 10 feet (3.1 m).

Each wheel had its own electric drive, a DC series-wound 0.25 horsepower motor capable of 10,000 rpm, and a mechanical brake unit. Power was provided by two 36-volt silver-zinc potassium hydroxide non-rechargeable batteries. Thermal controls kept the batteries within an optimal temperature range.

A T-shaped hand controller sat between the two seats. This controlled the four drive motors and two steering motors, along with the brakes. Pushing the stick forward sent the LRV forward, moving it to the left and right, steered it left and right, while pulling it backward applied the breaks. Operating a switch before pulling the stick back put it in reverse. Rovers were designed to have a top speed of 8 mph (13 kph), though on Apollo 17 Eugene Cernan managed 11.2 mph (18.0 kph).

The control and display modules sat in front of the handle, giving a read-out of speed, heading, pitch, and power and temperature levels. Navigation was based on continuously recording the direction and distance traveled through the use of a directional gyro. An onboard computer kept track of the distance and direction back to the LM. There was also a Sun-shadow device that could give a manual heading based on the direction of the Sun.

It carried color TV cameras, communication equipment and antenna, a secondary life-support system, equipment for drilling, and other tools for taking samples.

On Apollo 15 the LRV was driven a total of 17.3 miles (27.8 km) in 3 hours 2 minutes of drive time. The longest single traverse was 7.8 miles (12.5 km) and the maximum range from the LM was 3.1 miles (5.0 km). On Apollo 16 the vehicle traversed 16.6 miles (26.7 km) in 3 hours 26 minutes of driving. The longest single traverse was 7.2 miles (11.6 km) and the LRV reached a distance of 2.8 miles (4.5 km) from the LM. On Apollo 17 the rover went 22.3 miles (35.9 km) in 4 hours 26 minutes total drive time. The longest single traverse was 12.5 miles (20.1 km) and the greatest range from the LM was

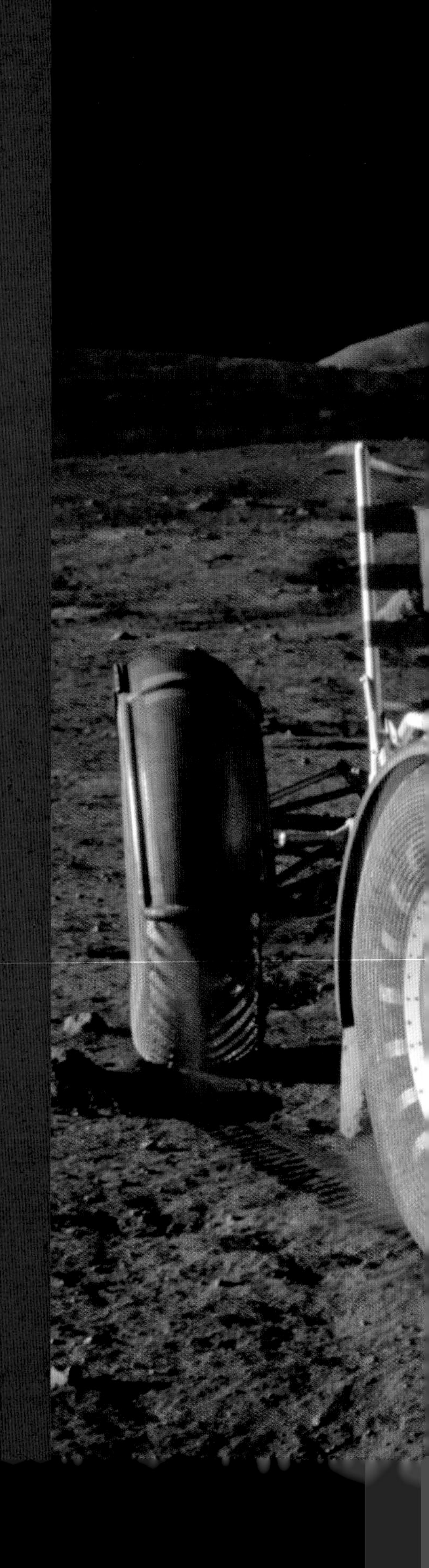

Eugene A. Cernan makes a short checkout of the Lunar Roving Vehicle (LRV) during the Apollo 17 extravehicular activity (EVA) at the Taurus-Littrow landing site.

Later Scott made the traditional landing on the Moon, walking backward down the ladder. Now he could be seen on TV. Recent improvements in color TV made the scene vivid. Turning to face the TV camera so his gold-tinted visor reflected the *Falcon* and the lunar horizon, he said:

> *As I stand out here in the wonders of the unknown at Hadley, I sort of realize there's a fundamental truth to our nature—Man must explore and this is exploration at its greatest.*

He then did a quick check of the spacecraft and found that the front footpad was not in contact with the surface, while the bell of the descent engine was buckled. Irwin also noticed that the front footpad was clear of the surface when he came down the ladder. As he put his weight on it, it turned, throwing him off balance.

"I would never have lived it down if I had fallen on my backside in full view of millions of television viewers," he said.

It could have been worse. If Scott had not turned the engine off so promptly, the back pressure from the new elongated bell on the LM risked blowing up not only the engine but also the entire spacecraft.

## LUNAR ROVING

The two men then set up a TV camera to record the deploying of the Lunar Rover. As they pulled the lanyards to draw it out of one of the compartments in the lander's octagonal descent stage, it tilted forward and gradually unfolded. Once it was sitting on the lunar dust, they locked the controls and seats into position.

Scott climbed onboard and put on his seatbelt—not an easy thing to do in a bulky spacesuit. Then he drove around the *Falcon*, putting the Moon buggy through its paces. He soon discovered that the steering on the front wheels was not working. However, the steering on the back wheels was and gave the vehicle sufficient maneuverability.

Parking the LRV next to the *Falcon*, they began loading it up with tools and equipment. A TV unit was mounted on the front with a high-gain gold-mesh antenna that had to be pointed directly at the Earth. The CapCom Joe Allen back in Houston said that the TV pictures of the scenery were breathtaking.

"Can't be half as breathtaking as the real thing though, Joe," said Scott. "Wish we had time to just stand here and look."

Soon they set off for Hadley Rille. Scott found the LRV hard to drive because the surface was so uneven it rocked and rolled. It was impossible to keep the hand-controller straight, so he had to concentrate on the surface directly in front of them, while Irwin could look further ahead to see where they were going.

---

*That is spectacular—the most beautiful thing I have ever seen.*

David R. Scott, Apollo 15
Mission Commander

---

## MOUNT HADLEY DELTA

Scott and Irwin stopped at Elbow Crater, where the rille made a sharp turn to the north, then they moved on to Mount Hadley Delta where a huge crater had been gouged out of the hillside. This was St. George Crater and they intended to collect boulders thrown out of it by the impact. However, on the slope below, they found only one. It was about 5 feet (1.5 m) in diameter.

As they got off the rover to sample it, Scott looked down the valley and said: "That is spectacular—the most beautiful thing I have ever seen."

Joe Allen in Houston then pointed out that the rock they were sampling was some 3.5 billion years old.

"Can you imagine that, Joe?" said Scott. "Here sits this rock, and it's been here since before creatures roamed the sea on our little Earth."

They then panned the camera around to show the rille that had rocks as big as houses in it.

"Tell me this isn't worth doing, boy!" said Scott.

However, they could not venture into the rille or St. George Crater, which had a gentle 20° slope. They might have been able to make their way down it either on foot or in the rover, but there may have been a problem getting back up. They were also under strict instruction not to venture so far from the LM that they could not walk back.

After 2 hours 15 minutes, they returned to the *Falcon*. They then had to deploy the ALSEP. Scott got out a battery-powered drill to bore two 10-foot (305-cm) holes for temperature probes to measure the heat escaping from the inside of the Moon in the hope of discovering whether the Moon had once had a molten core like that of the Earth. However, he could not drill down more than 5 feet (152 cm) as the subsurface was too hard and, in low-g, he could not put sufficient weight on it.

After 6 hours 32 minutes they clambered back into the LM to rest. Irwin was also badly dehydrated. The tube that carried water from the bag in his spacesuit had kinked and he had been unable to quench his thirst.

## GENESIS ROCK

The following day, they drove back to Mount Hadley Delta and took pictures of the *Falcon* which was 3 miles (4.8 km) away and 300 feet (91 m) below them. The front steering began working which made the LRV more maneuverable and they moved on, trying to find a crater where the surface of the Moon was broken and ancient material had been thrown out.

They moved on to Spur Crater where they saw a green rock in the distance and, thinking it may be significant, made for it. Parking below the boulder and parallel to the slope, they got out.

"As I moved toward the boulder, I saw the rover begin to slide down the side of the mountain," said Scott. "Our return to the LM was slipping away."

He alerted Irwin, who hopped down the slope and grabbed hold of the buggy. Scott took a sample of the boulder, then a photo of Irwin holding the rover with one wheel off the ground. After they moved on, they found green material on their boots too. The color came from tiny glass spheres formed in fire fountains where lava had been ejected from the ancient Moon.

Soon after they found a lump of white material the size of a fist glinting in the sunlight. This was the crystalline anorthosite they had been told to search for. The media called it "Genesis rock" which was around 4.15 billion years old—a fragment of the lunar crust thrown out by the impact that had made Spur Crater. Their geological training had paid off. Unfortunately their dwindling oxygen supplies prevented them searching further.

David R. Scott attempts to drill the lunar surface with the battery-powered drill.

## TORN MUSCLES AND DEHYDRATION

Back at the *Falcon,* Scott tried the drill again, but still failed to bore down more than 5 feet (152 cm). Trying a third time he managed to cut a 10-foot (305-cm) core sample, but it would have to wait until the next EVA before they could remove it. But then it took a great deal of heaving to prize it free. It was an important sample, showing fifty-eight layers of the Moon's history.

However, the effort had cost Scott a torn shoulder muscle and, on top of the dehydration the previous day, there were concerns about the strain on Irwin's heart. This meant that a trip to the North Complex had to be canceled. It was only 2 miles (3.2 km) north of the landing site and it was thought that the hills there were volcanic in origin.

Instead Scott and Irwin drove back to Hadley Rille where they were to take samples on the edge. On the exposed 1,000-foot (305-m) walls they could see layering, indicative of stratified lava beds. Bouncing down the slope, Scott suffered a fall and Irwin had to pull him up.

## THE FALLEN ASTRONAUT

Back at the LM they had to cut up the core sample so it could fit into the *Falcon*, then they loaded up the other samples, film, and equipment. Scott then gave a demonstration for TV viewers of an experiment first conducted by Galileo on the Leaning Tower of Pisa in the sixteenth century. In his left hand he held a falcon feather plucked from the Air Force Academy's mascot, in his right a geological hammer. He dropped them and they both hit the lunar dust at the same time.

> *Being on a roll, it was somewhat sad to leave our amazing and friendly lunar home.*
>
> David R. Scott, Apollo 15
> Mission Commander

Scott then parked the LRV 300 feet (91 m) to the east of the LM, so it would transmit back live TV pictures

Astronaut James B. Irwin gives a military salute while standing beside the deployed US flag during the Apollo 15 lunar surface extravehicular activity (EVA) at the Hadley-Apennine landing site.

of the LM's launch. Secretly he conducted a private ceremony paying tribute to the fourteen individuals who, by then, had lost their lives in the space program, either in training or in flight.

The most recent had been the three-man crew of Soyuz 11 which had depressurized during reentry, asphyxiating them. It had occurred just five weeks earlier.

Scott left a card carrying the names of the eight astronauts and six cosmonauts who had perished, along with a small figurine called "The Fallen Astronaut." Asked what he was doing while he was out of sight, Scott said: "Just cleaning up the back of the rover here a little, Joe."

## INTO THE WILD BLUE YONDER

After 2 days 18 hours on the Moon, Scott and Irwin blasted off in the *Falcon*. While they had been on the lunar surface, Worden had set the world's record for being the most isolated person ever when the *Endeavour*'s orbit took him 2,235 miles (3,597 km) from the nearest human beings. He had not been wasting his time though, conducting experiments and taking photographs that would immeasurably add to our knowledge of the Moon.

Particularly he was looking for signs of volcanic activity. He found them in the Taurus-Littrow area along the south-eastern edge of the Mare Serenitatis, or Sea of Serenity.

"I saw and described some small cone-shaped objects that appeared to be the results of volcanic eruptions," he said. Apollo 17 would be sent there to confirm his findings.

Worden saluted his returning colleagues by playing the US Air Force song "Off We Go Into The Wild Blue Yonder."

*The Moon is a pretty desolate body ... very wild looking but also very dead looking.*

Al Worden, Apollo 15
Command Module Pilot

## IRREGULAR HEART BEATS

When the *Falcon* docked back with *Endeavour*, the crew put on their pressure suits as a precaution before jettisoning the LM. Scott and Irwin had difficulty doing this due to lunar dust in the seals. Doctors at Mission Control spotted signs of dehydration and exhaustion. Irwin's heart beat became irregular. He was having a mild heart attack.

There was a further problem with dirt in the seal in the tunnel connecting the CSM and the LM, delaying separation by 2 hours 10 minutes. Then when the *Falcon* was released, the *Endeavour* had to make a small engine burn to avoid collision.

After seventy-four orbits of the Moon, Apollo 15 fired its main engine and started its three-day journey home. On the way Worden had to make a spacewalk to retrieve valuable film cassettes from the SIM. It was the first deep space EVA.

There were further problems with the splashdown when one of the parachutes collapsed during descent. Nevertheless the capsule and crew were recovered safely by USS *Okinawa*. Their mission had lasted 12 days 7 hours 12 minutes.

## Lunar Scandal

Two hundred and fifty postal covers were officially taken to the Moon and back on Apollo 15 for the US Postal Service. However, the astronauts carried 400 extra. One hundred each would be for their own use. They were given this concession as it was impossible for them to take out life insurance. The other hundred were to be given to a German stamp dealer. In return, he would set up a $6,000 trust fund for each of the astronauts for the education of their children. Under the terms of the agreement, he was not to sell the covers until the end of the Apollo program. However he did, and the resulting scandal meant the astronauts had to return the money. They were reprimanded for bad judgment by the Senate Space Committee.

The figurine of "The Fallen Astronaut" Scott left on the Moon was also the subject of controversy after its existence was revealed in a post-flight press conference. The Smithsonian then wanted a copy and the sculptor Paul Van Hoeydonck wanted credit for his work. He agreed to make 950 replicas, but was prevailed on not to by NASA who had a strict policy against commercial exploitation of the space program.

# CHAPTER 22

# JOURNEY TO THE LUNAR HIGHLANDS

## APOLLO 16

Other Apollo missions had headed for the darker areas of the lunar landscape known as seas. Instead Apollo 16, the fifth and penultimate mission to land on the Moon, was heading for the lunar highlands which appeared brighter. It would make the most southerly landing of the Apollo program, nine degrees south of the lunar equator near a crater named Descartes (after the French philosopher and mathematician). The site had been selected from photographs taken by the Command Module pilot of Apollo 14 Stuart Roosa.

The Commander on the mission was 41-year old Navy Captain John Young. Before joining NASA in 1962 as part of its second nine-man astronaut team, he had been a test pilot who held two world altitude climb records. He had flown the first Gemini mission with Gus Grissom and commanded Gemini 10 with Mike Collins. Then he served as Command Module Pilot on Apollo 10. On Apollo 16 he would get his chance to stand on the Moon. The Lunar Module Pilot of Apollo 16 was 36-year-old Air Force Lieutenant Colonel Charles Duke who had been the CapCom on Apollo 11. The Command Module Pilot was 36-year-old Thomas "Ken" Mattingly who had been slated for Apollo 13, but was dropped when it was discovered that he had been exposed to German measles.

*I'm not normally a rabble rouser, but something funny is going on here.*

John Young, Apollo 16 Mission Commander

## BROWN DEBRIS

Apollo 16's launch was delayed by a month due to technical problems. The launch took place on April 16, 1972, from Launch Complex 39 at the Kennedy Space Center in Florida. It went off smoothly, but when it was 45,000 miles (72,420 km) from Earth a stream of flaky brown debris was seen coming off the LM *Orion*. Mattingly said it was like paint from an old barn, while Duke compared it to shredded wheat.

Young and Duke entered the LM to check out the systems, but could find nothing wrong. Grumman Aircraft, who built the LM, soon identified the flaking material as thermal insulation paint which had been applied to protect the LM from the heat of the Sun. Shedding it would not endanger the astronauts or the mission, provided they used the usual "rotisserie" technique, which was to rotate the spacecraft slowly to even out the heating.

They were 139,000 miles (223,698 km) from Earth when there was a problem with the main guidance system on the CSM *Casper*. This was a unit with a suspended navigation platform, controlled by a gyroscope. A "gimbal lock" warning light lit up indicating that it was not reporting the spacecraft's attitude. The platform was stuck in a single position and not responding to the movements of the CSM. Mission Control had noticed this too.

There was, of course, a backup in the LM. However, if they had to depend on that the two craft could not be separated and the Moon landing would have to be canceled. The Guidance Officer in Houston worked out how to free the CSM's gyroscope and radioed instructions up to Mattingly.

To realign the inertial platform, Mattingly would have to take sightings of the stars. This proved impossible as flakes of thermal insulation paint were still traveling alongside. Instead he took sightings of the Sun, which stabilized the platform. Gradually, as he became able to take accurate star sightings, he fine-tuned the alignment and, after forty-five minutes, the guidance system was working properly again. Young and Duke had slept through this potential crisis, so Mattingly wrote up what he had done, leaving a note for the other two before he too went to sleep.

## A MINUTE TO SPARE

Young and Duke were suited up and in the LM starting up the system ready for separation when another problem occurred. When he powered up the primary antenna Duke found he could not rotate to point in the direction of Earth. This meant they were not able to uplink information to their computer automatically. The data would have to be entered into *Orion*'s computer manually. This was time-consuming and

Thomas K. Mattingly (right) performs an extravehicular activity (EVA) during the Apollo 16 trans-Earth coast, assisted by astronaut Charles M. Duke Jr.

had to be done before they went behind the Moon and lost radio contact. They did it with one minute to spare. Having met the deadline, the CSM and the LM could separate, leaving *Orion* free to make its landing.

Duke had done this against all the odds. His EVA suit was uncomfortable, but Houston refused permission to loosen the laces. Worse still, juice was squirting from the plastic straw inside his spacesuit, designed to sustain the astronauts when they were on the lunar surface. The liquid was also smearing his visor, making it difficult to see. Young complained to Mission Control about this using some choice language, only to be warned by Houston that the transmission was being broadcast worldwide.

## SHAKING THE SPACECRAFT TO PIECES

When the two craft separated behind the Moon, Mattingly was to fire the SPS engine making the orbit circular at about 60 miles (100 km) above the Moon. Suddenly, he cried out: "There's something wrong with the secondary control system to the engine. When I turn it on, it feels as though it is shaking the spacecraft to pieces."

Only the two men on *Orion* could hear this and they knew that *Casper* was their only way back to Earth. Young ordered Mattingly to delay the maneuver, so that when they emerged from behind the Moon they could report the fault to Houston. At the same time, Mission Control was agonizing over whether to give Young permission to land *Orion* on the lunar surface.

When they emerged from behind the Moon, Young was ordered to stay close to *Casper*. If necessary they could redock and use the engines of the LM to take them home. It seemed to flight controllers that they were reliving Apollo 13. The decision was taken to make five more orbits around the Moon. In the 7 hours 30 minutes this would take, technicians from North American Rockwell, who had built the Command Module, would try and get the problem sorted out.

## YAW GIMBAL

Finally, the problem was identified. It concerned the yaw gimbal that moved the bell of the engine left and right. It was not working due to an open circuit in the backup guidance system. Even so the backup system could still be used if the primary system failed, so the mission could proceed and the go-ahead for the lunar landing was given.

*Orion* came in a bit steep, but Duke could not contain his excitement.

"Coming in like gangbusters," he said. "I can see the landing site from here. Go at eight. John's got a visual. We're right on John. Right back on profile. Right on profile ..."

They came in fast over the craters Gator and Lonestar that they recognized from the 3D images they had seen on the flight simulator.

"Stand by for contact. I'm going to let her down ... level off. Let her on down ... Contact! Stop. Whew ... WOW! ... Well we don't have to walk far to pick up rocks."

To conserve energy due to the longer time they had spent in lunar orbit, they powered down immediately. The disadvantage of this was that it also cut the mission timer that provided emergency liftoff times if there was an emergency. So if they lost communication with Earth, they would have to use their wristwatches to work out when to blast off to rendezvous with *Casper*.

---

*When we did it for the first time, I mean for real, in flight ... it looked exactly like the mock-up.*

Charles Duke, Apollo 16
Lunar Module Pilot

---

## MYSTERIOUS AND UNKNOWN

After a quick breakfast of cubes of cinnamon toast, peaches, and cold scrambled eggs, they were ready for their first EVA. Young descended the ladder, saying: "There you are, our mysterious and unknown Descartes highland plains. Apollo 16 is going to change your image."

Soon after Young set foot on the surface he got the news that Congress had passed the bill giving the go-ahead to the Space Shuttle. Nine years later he would command the shuttle's maiden flight.

The flag was erected. The LRV was brought out and the ALSEP was deployed. As with Apollo 15 a 10-feet-deep (305-cm) hole had to be drilled to ascertain the amount of heat that was flowing out from the interior of the Moon. The drill had been upgraded and Duke managed this in a few minutes. Then the thermometer was inserted. However, Young tripped on one of the experiment's cables and accidentally disconnected it. There was no time to make repairs.

## A GLINT FROM THE LRV

Young took the LRV for a spin, while Duke filmed what they called the "Descartes Rally." With the low-g and the rough lunar surface, it looked like Young barely had the wheels on the ground at all.

"Indy's never seen a driver like this," said Duke.

They headed 4 miles (6.4 km) south over the 1,800-feet (5,486 m) Stone Mountain and looked out over the valley they had landed in. In the middle they could see the LM, while from space Mattingly reported seeing a glint from the LRV itself.

Orbiting 60 miles (97 km) up, Mattingly had plenty to occupy himself with. He, too, had been given geological training by Lee Silver and Farouk El-Baz. He had even learned how to work out the slope of a hill by the amount of small rocks, or talus, at its foot. Otherwise he listened to Gustav Holst's *The Planets* while watching Earthrise.

On another EVA, Young and Duke hopped over to a large rock they could see from the LM. With few clues about perspective, it was hard to tell whether it was an average-sized rock relatively close, or a huge rock far away. It turned out to be massive.

They named it "House Rock" and took samples. In all Apollo 16 returned with 215 pounds (97.6 kg) of rocks and dust—the biggest haul so far. However, most of it was anorthosite and breccias—rock consisting of angular fragments cemented together—which had been picked up in previous missions. This told geologists back on Earth that the surface of the Moon had largely been formed by cosmic bombardment, rather than volcanic action as they had thought.

The Lunar Roving Vehicle (LRV) gets a speed workout by astronaut John W. Young in the "Grand Prix" run during the first Apollo 16 extravehicular activity (EVA) at the Descartes landing site.

## LUNAR OLYMPICS

Before they left the Moon, Young and Duke staged the "Lunar Olympics" for the TV audience. Duke was determined to set a new high jump record, but toppled over.

"I was in trouble," he said. "You could watch me scrambling, trying to get my balance. I ended up landing on my right side and bouncing onto my back. My heart was just pounding."

Young helped him to his feet. The danger was that he could have punctured his suit.

"Charlie, that ain't very smart," Young said.

They had spent twenty hours exploring the lunar surface and seventy-one hours after landing they blasted off. After rendezvousing with *Casper*, the crew gave the two craft a quick inspection. To avoid the problems with lunar dust encountered on Apollo 15, Young and Duke wiped themselves down with damp towels and when they opened the hatch they were greeted by Mattingly brandishing a vacuum cleaner.

This picture from a color television transmission made by a camera mounted on the Lunar Roving Vehicle (LRV) shows the early lunar liftoff phase of the Apollo 16 Lunar Module at the Descartes landing site.

## WHEN THERE ARE NO STARS

Once the samples and equipment had been moved from *Orion* into *Casper*, the LM was jettisoned. This immediately proved a danger. Because of a switch left in the wrong position, it began to tumble uncontrollably and Mattingly quickly had to take evasive action to avoid a collision.

During the journey home, Mattingly had to make a spacewalk to retrieve the film cassettes from the SIM bay. They were 180,000 miles (289,628 km) from Earth, traveling at 5,000 mph (8,000 kph) when he clambered over the outside of the spacecraft. Duke also popped out of the hatch to expose various biological samples to cosmic rays.

During the EVA Mattingly was puzzled that he could not see any stars. The Earth was out of sight and only the Moon was visible. He realized that this was because of his gold-tinted visor. When he lifted it, the heavens were full of points of light. He said:

> *That's the only time that I had a sensation of being away. I don't know why, but around the Moon, inside the spacecraft, I didn't ever have a sense of being that far away until I looked out there when there were no stars and the entire world was within 15 feet (4.5 m), and there was nothing else.*

## THE BAD SMELL OF SPACE

In the Pacific Ocean a US Navy helicopter from USS *Ticonderoga* was waiting for them to splashdown. Mattingly had trouble opening the hatch which was slammed back in his face by one of the Navy rescue divers. Later the diver took him back to the capsule to explain what had happened. When Mattingly opened the hatch the bad smell coming out of the capsule was unbearable.

"The odor was unbelievable," Mattingly said. "Thank God our olfactory nerves disappear early in life. Living in a bathroom for ten days is not a desired position. *Life* magazine doesn't tell you about those things."

Mattingly went back into space as commander of the Space Shuttle *Columbia* in 1982 and *Discovery* in 1985.

# CHAPTER 23

# LAST OF THE MOON MISSIONS

## APOLLO 17

Apollo 17, the last Moon mission, blasted off on December 7, 1972. The countdown was halted thirty seconds before its scheduled liftoff, by a computer malfunction, which caused a delay of 2 hours 40 minutes with the crew sitting patiently on top of the rocket.

The Commander was Gene Cernan who had flown twice before. The other two crewmen were rookies. The Command Module pilot was Ron Evans and the Lunar Module pilot Dr. Harrison "Jack" Schmitt, a geologist, who replaced veteran X-15 pilot Joe Engle at the last moment. Pressure had been applied by Washington to include a scientist and Schmitt had helped develop the ALSEP. He had also learned to fly after being selected for the astronauts corps and had been involved in selecting the landing site for Apollo 17.

The objective was the Taurus-Littrow valley where Al Worden had spotted signs of volcanic activity on Apollo 15. This offered the opportunity to discover when the Moon was last volcanically active.

Schmitt was not just interested in rocks. He also took a scientific interest in meteorology and gave a running commentary on the weather patterns below. Sometimes he was drowned out by an alarm that first sounded in Earth orbit, then intermittently on the outward journey until the problem was resolved.

*During each staging the fireball overtook us—then when the engine kicked in we once again flew out of the orange-red cloud into darkness.*

Ron Evans, Apollo 17
Command Module Pilot

## NO CUTTING EDGE

On the way to the Moon, the CM *America* and the LM *Challenger* transposed without a hitch. The only slight problem was that Evans lost the pair of scissors he needed to open food packages.

Cernan and Schmitt had to leave one of their pairs with him so that he could eat while they were on the lunar surface. Cernan described the descent:

> *All the way down it's noisy; it's vibrating. We're in our suits. It's a very dynamic period of time. The ground's talking to you. Guy in the right seat, in my case Jack Schmitt was talking to me. I'm flying. A lot of people think we pressed a button and let the thing fly itself … Nobody ever landed on the Moon other than with their own two hands and brain and eyeballs and whatever. Computer-assisted, yes. Got a lot of information. We got help from a lot of sources. But you're looking for landing radar. You're looking to maintain the communications. You're on your back. You've got to roll over. You've got to go face up. A lot of things happen very quickly. As I say a very dynamic, exciting fourteen minutes of your life, maybe fifteen. At 7,000 feet you pitch over, so for the first time you can really see the landing site where you're going to land.*

## CUTTING IT FINE

Landing in the the Taurus-Littrow valley also proved a challenge as Schmitt recalled later after the mission:

> *Our valley where we were to land in was surrounded by mountains on three sides that are higher than the Grand Canyon is deep, to give you some idea. So at 7,000 feet we were down among them. I mean the mountains rose above us on both sides. The valley was only twenty miles long and about five miles wide.*

*Challenger* set down on Taurus-Littrow on December 11, 1972, only 200 feet (60 m) from their intended landing site. They were cutting it fine with only seventeen seconds of fuel left, prompting Schmitt to say: "We should have hovered around a little bit … gone and looked at that scarp."

As a rest before the first EVA had been foregone on earlier missions, Cernan and Schmitt left the LM shortly after touchdown. Schmitt promptly tripped and dropped their only pair of scissors which sank into the lunar dust. Fortunately, a couple of minutes later, he found them.

Setting up the ALSEP experiments took longer than planned. This ate into the schedule which had originally involved investigating ten geological sites over their three EVAs which were each to last for seven hours. They took over 2,000 photographs and collected 250 pounds (113 kg) of rock and dust.

## ORANGE SOIL

During the first EVA, Cernan snagged the LRV's fender with a geological hammer and tore it off. He tried mending it with duct tape, but because of the lunar dust this would not hold. Later Houston

Eugene A. Cernan eating a meal under weightless conditions of space during the final lunar landing mission.

Astronaut Eugene A. Cernan approaches the parked Lunar Roving Vehicle (LRV) on the lunar surface during the flight's third period of extravehicular activity (EVA-3).

sent instructions to fix it using sturdy geological maps before exploring further. Schmitt later described Taurus-Littrow:

> *We had three-dimensions to look at with the mountains, to sample. You had the Mare basalts in the floor and the highlands in the mountain walls. We also had this apparent young volcanic material that had been seen on the photographs and wasn't immediate obvious, but ultimately we found in the form of the orange soil at Shorty crater.*

High above, Evans took over 5,000 high-resolution photographs, made a detailed temperature map of the lunar surface, and took measurements of the thin lunar atmosphere. After Cernan and Schmitt had discovered orange soil on the surface, Evans was tasked with seeing if he could find other patches from overhead. He spotted small patches, but had no time to make a comprehensive survey.

At first, this was thought to suggest recent volcanic activity. Later it was found that the color was due to zircon-rich glass beads that had been exposed by meteorite impact. It was around four billion years old, as old as the basaltic rock surrounding it.

## TWINKLE TOES

While Cernan used the traditional "bunny hop" to get around, Schmitt who had spent time in Norway got about using a method adopted from cross-country skiing.

"You could go six, ten kilometers an hour quite easily," he said. "If I had had a couple of ski poles, I could probably have got up to fifteen and held that for some time. I think I could have outrun the rover."

This earned him the nickname "Twinkle Toes" though he frequently fell. CapCom Bob Parker joked: "Be advised that the switchboard here at MC has been lit up with callers from the Houston Ballet Foundation requesting your services for next season."

Cernan said that walking on the Moon was painful for him as he had injured a tendon in his legs two months earlier playing softball. He said nothing about it, fearing he would be dropped from the mission.

On the day before returning to the CM, Cernan drove the rover to a point away from the LM so that a camera on the vehicle could film their departure. He then paid tribute to his young daughter, Tracy.

"I took a moment to kneel and with a single finger, scratched Tracy's initials, TDC, in the lunar dust, knowing those three letters would remain there undisturbed for more years than anyone could imagine," Cernan wrote in his memoir *The Last Man on the Moon*.

Returning to the LM after the last EVA, they took some more gravity measurements and gathered samples of the exhaust residuals from the descent engines. Then they loaded their rock samples onto the *Challenger*.

## LAST LUNAR WORDS

Schmitt climbed onboard and began brushing off the lunar dust, while Cernan made a short speech marking the end of the Apollo program. As he lifted his foot from the lunar soil for the last time, he said:

> *As I take man's last step from the surface … I'd like to just say what I believe history will record. That America's challenge of today has forged man's destiny of tomorrow. And, as we leave the Moon at Taurus-Littrow, we leave as we came and, God willing, as we shall return, with peace and hope for all mankind. Godspeed the crew of Apollo 17.*

Reflecting on his visit to the Moon, he said later: "I knew that I had changed in the past three days and that I no longer belonged solely to the Earth."

During their stay on lunar surface, the pair of astronauts had clocked up over twenty-two hours of EVA, and traveled more than 19 miles (30 km) in the lunar rover.

Assessing the achievements of the mission, Cernan said:

> *Apollo 17 built upon all of the other missions scientifically. We had a Lunar Rover, we were able to cover more ground than most of the other missions. We stayed there a little bit longer. We went to a more challenging unique area in the mountains, to learn something about the history and the origin of the Moon itself.*

As they prepared for liftoff from the surface, Schmitt was fiddling with a camera. Annoyed, Cernan suggested Schmitt get out and film the launch.

"OK, now let's get off," he snapped. "Forget the camera."

The last words spoken on the lunar surface went to Schmitt. They were: "Three, two, one—ignition."

## THE END OF THE BEGINNING

There was then a four-minute loss of communication between *Challenger* and Houston as the engine plume generated too much static. Once *Challenger* had docked with *America*, the spacecraft stayed in orbit for two more days, completing the orbital observation program.

On the way back to Earth, Evans took his time retrieving the three film canisters from the SIM bay. Cernan had suffered exhaustion on his own spacewalk during the Gemini 9 mission and told Evans: "When you get out there, just take it nice and easy. You've got all day long."

Schmitt added: "Nice day for an EVA, Ron. Go out there and have a good time."

Evans was out of the spacecraft for sixty-six minutes and he certainly enjoyed himself. He mounted a camera on a pole so that he could wave to his family as he chuckled and hummed.

"Hey, this is great," he said. "Talk about being a spaceman."

The Apollo program ended with the splashdown of Apollo 17, 400 miles south-east of Samoa, and its recovery by USS *Ticonderoga* on December 19, 1972. According to Cernan, the crew quickly got tired of hearing their mission called "the end." He said in January 1973:

> *We are not the end. We are just the end of the beginning. Apollo is the beginning of a whole new era in the history of mankind. Not only are we going back to the Moon, but we are going to be on our way to Mars by the turn of the century.*

Unfortunately Gene Cernan was wrong, and no humans have set foot on the Moon since December 1972.

Harrison Schmitt adjusts the last American flag to be planted on the Moon before the Apollo 17 mission made their homeward journey to the Earth.

# THE BLUE MARBLE

On their way to the Moon, the crew of Apollo 17 took one of the most iconic photographs in space-program history, the full view of the Earth dubbed "The Blue Marble." Despite its fame, Cernan felt the photograph had not really been appreciated. In 2007, he explained:

> *What is the real meaning of seeing this picture? I've always said, I've said for a long time, I still believe it, it's going to be—well it's almost fifty now, but fifty or a hundred years in the history of mankind before we look back and really understand the meaning of Apollo. Really understand what humankind had done when we left, when we truly left this planet, we're able to call another body in this universe our home. We did it way too early considering what we're doing now in space. It's almost as if JFK reached out into the twenty-first century where we are today, grabbed hold of a decade of time, slipped it neatly into the [19] 60s and 70s [and] called it Apollo.*

View of the Earth as seen by the Apollo 17 crew traveling toward the Moon. This translunar coast photograph extends from the Mediterranean Sea area to the Antarctica South polar ice cap. This is the first time the Apollo trajectory made it possible to photograph the South polar ice cap. Almost the entire coastline of Africa is clearly visible. The Arabian Peninsula can be seen at the north-eastern edge of Africa. The large island off the coast of Africa is Madagascar. The Asian mainland is on the horizon toward the north-east.

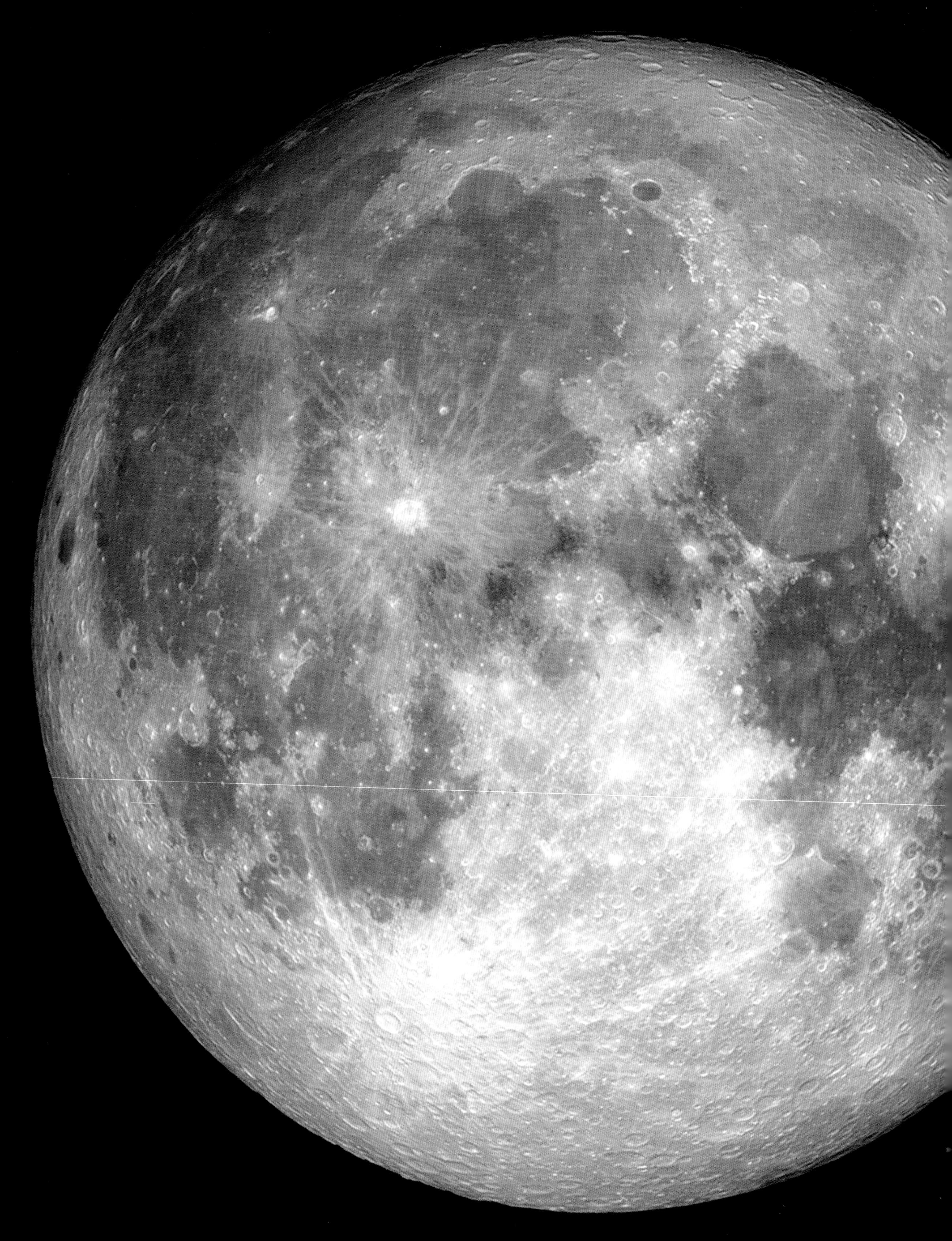

# FURTHER READING

This book documents NASA's Apollo Space Program from 1961 to 1972 and is designed to be an informative and entertaining introductory text. There are many more academic publications available should the reader wish to delve more deeply. Publications that were especially useful during the preparation of this book are listed below, and other credits are cited at the point where they appear within the text.


Aldrin, Buzz (1989) *Men From Earth*. Bantam, New York.

Burgess, Colin (2010) *Footprints in the Dust: The Epic Voyages of Apollo, 1969–1975*. University of Nebraska Press, Omaha.

Cernan, Eugene (1999) *Last Man on the Moon*. St. Martin's Press, New York.

Collins, Michael (2009) *Carrying the Fire*. Farrar, Straus and Giroux, New York.

Chaikin, Andrew (1994) *A Man on the Moon: The Voyages of the Apollo Astronauts*. Penguin Books, New York.

Floca, Brian (2009) *Moonshot: The Flight of Apollo 11*. Atheneum/Richard Jackson Books, New York.

French, Francis & Burgess, Colin (2007) *In the Shadow of the Moon: A Challenging Journey to Tranquility 1965–1969*. University of Nebraska Press, Omaha.

Harland, D.M. (2008) *Exploring the Moon: The Apollo Expeditions*. Springer-Praxis, Chichester, UK.

Kranz, Gene (2009) *Failure Is Not An Option*. Simon & Schuster, New York.

Lovell, James & Kluger, Jeffery (1994) *Lost Moon: The Perilous Voyage of Apollo 13*. Houghton Mifflin, Boston.

Riley, Christopher & Dolling, Philip (2010) *Apollo 11: Owners' Workshop Manual*. Haynes Publishing, Somerset, UK.

Riley, Christopher, Woods, David & Dolling, Philip (2012) *Lunar Rover Manual: 1971-1972 (Apollo 15-17; LRV1-3 & 1G Trainer)*. Haynes Publishing, Somerset, UK.

Shepard, Alan B.(1994) *Moon Shot: Inside Story of America's Race to the Moon*. Virgin, London.

Von Braun, Wernher & Whipple, Fred L. (1953) *Conquest of the Moon*. Viking Press, New York.

Wolfe, Tom (2004) *The Right Stuff*. Black Dog & Leventhal, New York.

Woods, David (2016) *NASA Saturn V 1967–1973 (Apollo 4 – 17) Owners' Workshop Manual*. Haynes Publishing, Somerset, UK.

# INDEX

Note: Page numbers in *italic* refer to photographs

Brimming with creative inspiration, how-to projects, and useful information to enrich your everyday life, Quarto Knows is a favorite destination for those pursuing their interests and passions. Visit our site and dig deeper with our books into your area of interest: Quarto Creates, Quarto Cooks, Quarto Homes, Quarto Lives, Quarto Drives, Quarto Explores, Quarto Gifts, or Quarto Kids.



This edition published in 2019 by Chartwell Books,
an imprint of The Quarto Group,
142 West 36th Street, 4th Floor,
New York, NY 10018, USA
**T** (212) 779-4972 **F** (212) 779-6058
**www.QuartoKnows.com**



10 9 8 7 6 5 4 3 2

ISBN: 978-0-7858-3703-9

Printed in China

## Picture Credits

We are very grateful to the NASA Apollo Image Archive (nasa.gov), the Kennedy Space Center (KSC), and the Marshall Space Flight Center's Marshall Image Exchange (MSFC) who provided the images in this book apart from those credited below. Interested readers can visit their websites for further information. The other image contributors are:

8 Photo 12 / Alamy / 10 Leon Werdinger / Alamy / 11 Diamond Images / Getty / 12 Bob Gomel / The Life Images Collection / Getty / 14 Arturas Medvedevas / 15 Hulton Archive / Getty / 17 Sputnik / Alamy / 18 Science History Images / Alamy / 21 Rolls Press / Popperfoto / Getty / 22 State Archives of Florida / Florida Memory / Alamy / 24 Horyzonty Techniki 11/1966 / 25 Military Collection / Alamy / 30 SSPL / Getty / 32 National Reconnaissance Office (NRO) / 36 Keystone Pictures USA / Alamy / 38 left NASA Image Collection / Alamy / 38 middle NASA Archive / Alamy / 41 top Heritage Image Partnership Ltd / Alamy / 41 bottom Sputnik / Alamy / 43 Interfoto / Alamy / 44 jfklibrary.org / 47 strangerplanets.com / 48 rawpixel.com / 52 Ralph Morse / The Life Picture Collection/Getty / 53 Science History Images / Alamy / 54 RGB Ventures / SuperStock / Alamy / 55 top NASA Photo / Alamy / 57 NASA Image Collection / Alamy / 67 orenda.pl.ua / 91 Photo 12 / Alamy / 96 Sputnik / Alamy / 115 NASA Photo / Alamy / 131 Art Directors & Trip / Alamy / 145 NASA Photo / Alamy / 157 NASA Image Collection / Alamy / 164 Apollo Photo Archive Collection / Alamy / 173 NASA Image Collection / Alamy

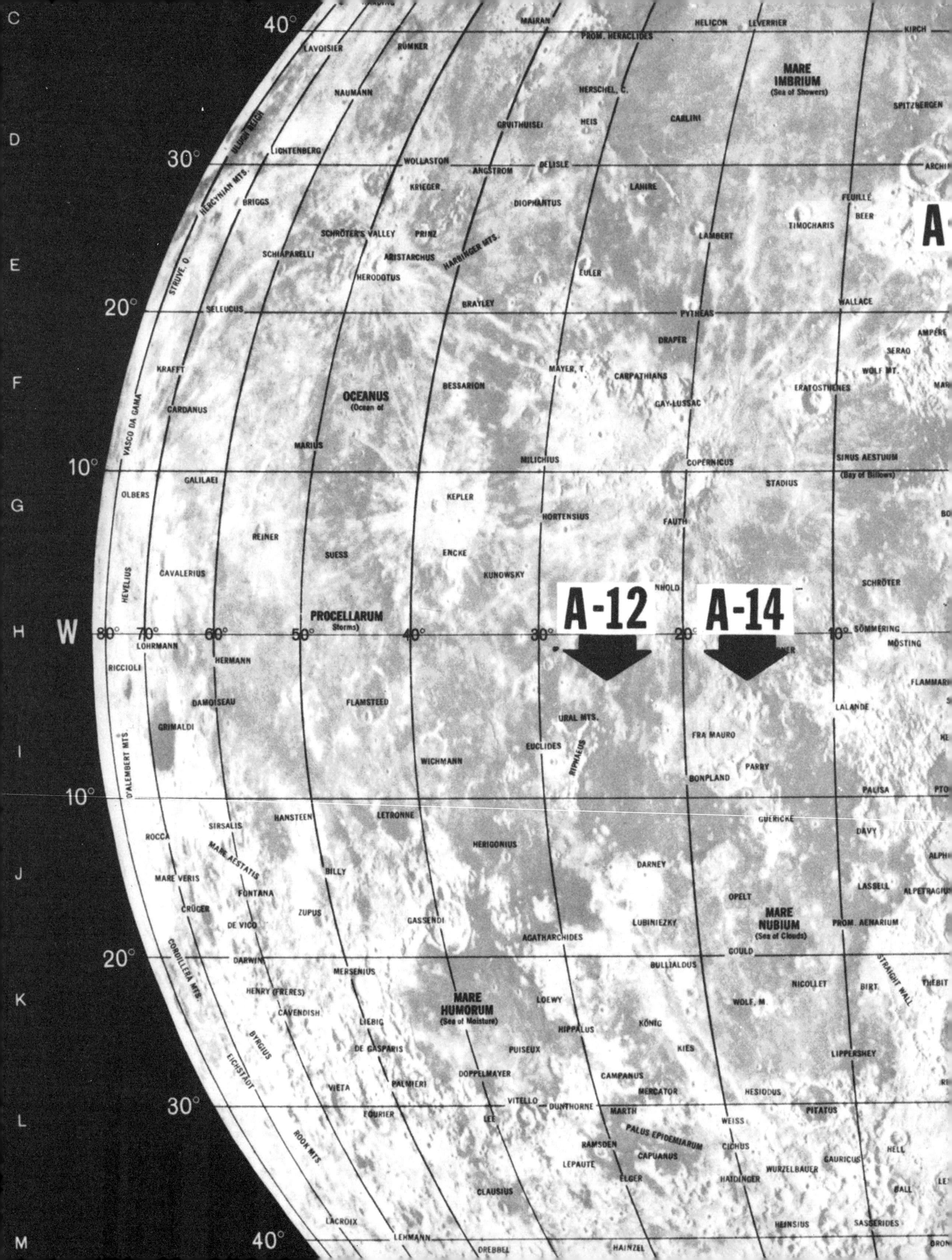

C
D
E
F
G
H
I
J
K
L
M
W
40°
30°
20°
10°
10°
20°
30°
40°
80°
70°
60°
50°
40°
30°
20°
10°
A-12
A-14
MAIRAN
HELICON
LEVERRIER
KIRCH
PROM. HERACLIDES
LAVOISIER
RÜMKER
MARE IMBRIUM
(Sea of Showers)
NAUMANN
HERSCHEL, C.
SPITZBERGEN
ULUGH BEIGH
GRUITHUISEN
HEIS
CARLINI
LICHTENBERG
WOLLASTON
ANGSTROM
DELISLE
HERCYNIAN MTS.
KRIEGER
LAHIRE
BRIGGS
DIOPHANTUS
FEUILLÉ
BEER
SCHRÖTER'S VALLEY
PRINZ
LAMBERT
TIMOCHARIS
SCHIAPARELLI
ARISTARCHUS
HARBINGER MTS.
STRUVE, O.
HERODOTUS
EULER
SELEUCUS
BRAYLEY
WALLACE
PYTHEAS
AMPÈRE
DRAPER
SERAO
KRAFFT
MAYER, T.
CARPATHIANS
WOLF MT.
BESSARION
ERATOSTHENES
OCEANUS
(Ocean of
VASCO DA GAMA
CARDANUS
GAY-LUSSAC
MARIUS
MILICHIUS
SINUS AESTUUM
COPERNICUS
(Bay of Billows)
GALILAEI
STADIUS
OLBERS
KEPLER
HORTENSIUS
FAUTH
REINER
SUESS
ENCKE
CAVALERIUS
KUNOWSKY
SCHRÖTER
HEVELIUS
PROCELLARUM
Storms)
SÖMMERING
LOHRMANN
MÖSTING
RICCIOLI
HERMANN
FLAMSTEED
DAMOISEAU
LALANDE
URAL MTS.
GRIMALDI
FRA MAURO
EUCLIDES
D'ALEMBERT MTS.
RIPHAEUS
WICHMANN
PARRY
BONPLAND
PALISA
SIRSALIS
HANSTEEN
LETRONNE
GUERICKE
ROCCA
DAVY
HERIGONIUS
MARE AESTATIS
DARNEY
BILLY
MARE VERIS
LASSELL
FONTANA
OPELT
CRÜGER
ZUPUS
MARE NUBIUM
(Sea of Clouds)
DE VICO
GASSENDI
LUBINIEZKY
PROM. AENARIUM
AGATHARCHIDES
CORDILLERA MTS.
GOULD
DARWIN
BULLIALDUS
MERSENIUS
STRAIGHT WALL
HENRY (FRÈRES)
NICOLLET
BIRT
THEBIT
MARE HUMORUM
(Sea of Moisture)
LOEWY
CAVENDISH
WOLF, M.
LIEBIG
KÖNIG
BYRGIUS
HIPPALUS
DE GASPARIS
KIES
PUISEUX
LIPPERSHEY
EICHSTADT
VIETA
PALMIERI
DOPPELMAYER
CAMPANUS
MERCATOR
HESIODUS
VITELLO
DUNTHORNE
MARTH
PITATUS
FOURIER
LEE
WEISS
PALUS EPIDEMIARUM
ROOK MTS.
RAMSDEN
CICHUS
HELL
CAPUANUS
GAURICUS
LEPAUTE
ELGER
HAIDINGER
WURZELBAUER
CLAUSIUS
BALL
LACROIX
HEINSIUS
SASSERIDES
LEHMANN
DREBBEL
HAINZEL